D1469784

GETTING TO KNOW

ArcObjects

Programming ArcGIS® with VBA

For ESRI® ArcView®, ArcEditor™, and ArcInfo™

Robert Burke

ESRI PRESS

REDLANDS, CALIFORNIA

ESRI Press, 380 New York Street, Redlands, California 92373-8100

Printed in the United States of America

Library of Congress Cataloging-in-Publication Data
Burke, Robert, 1963–
Getting to know arcobjects : programming ArcGIS with VBA / Robert Burke.
        p.   cm.
    ISBN 1-58948-018-X (pbk. : alk. paper)
    1. Geographic information systems. 2. ArcGIS. 3. Microsoft Visual Basic for applications.
    4. Graphical user interfaces (Computer systems) I. Title.
    G70.212.B86 2003
    910'.285'536—dc22   2003019217

Ask for ESRI Press titles at your local bookstore or order by calling 1-800-447-9778. You can also shop online at www.esri.com/esripress. Outside the United States, contact your local ESRI distributor.

ESRI Press titles are distributed to the trade by the following:

In North America:
Ingram Publisher Services
Toll-free telephone: (800) 648-3104
Toll-free fax: (800) 838-1149
E-mail: customerservice@ingrampublisherservices.com

In the United Kingdom, Europe, and the Middle East:
Transatlantic Publishers Group Ltd.
Telephone: 44 20 7373 2515
Fax: 44 20 7244 1018
E-mail: richard@tpgltd.co.uk

Cover design by Doug Huibregtse
Interior design by Michael Hyatt

ACKNOWLEDGMENTS

This book was crafted by a team of dedicated and talented individuals. Without their skill, patience, and expertise, *Getting to Know ArcObjects* would only be an idea.

Thad Tilton and Andrew Arana gave the book its initial push. They helped design it and plan its exercise scenarios, and they wrote early drafts of several chapters. Helpful advice came from Gary Amdahl, Michael Zeiler, Tom Gross, Euan Cameron, Jack Horton, Glenn Meister, Matt Crowder, Jerry Sommerfeld, and Brian Goldin.

Several rounds of edits by Michael Karman created the book's logical flow. Tim Ormsby helped write the final draft and created the index.

Tom Brenneman, Brian Parr, Tim Ormsby, Sheila Ferguson, Janis Davis, Sally Swenson, Tom Hurst, Jovanna Leonardo, Brandon Whitehead, Tiffany Modlin, and Michael Waltuch reviewed, tested, and technically edited every exercise. Brian Parr took care of the behind-the-scenes details, making sure each milestone was met in order to finish on schedule.

A special thanks to Doug Huibregtse for creating such an engaging book cover; to Michael Hyatt for book design, production, and copyediting; and to Steve Hegle for handling any and all administrative and distribution issues.

For their support and daily heroic efforts, I am grateful to my wife Suzanne and sons Ryan and Ethan.

And finally, my deep appreciation to Christian Harder, Judy Boyd, Nick Frunzi, and Jack Dangermond for helping me carry out a dream.

ArcGIS Desktop has evolved and improved a lot since this book's initial publication. ESRI has created two new specialized developer products called ArcGIS Engine for developing stand-alone GIS applications and ArcGIS Server for developing GIS Web applications. All three products share the same underlying ArcObjects technology.

ArcObjects are developed using the COM (Component Object Model) style of programming. One of COM's basic rules is that after an object (or class) and its interfaces have been published they will not change over time. For this book's ArcGIS 9.2 update, none of the original code samples had to be changed even though some of them were written in 1999.

Have fun learning about ArcObjects.

Robert Burke
December 2006

Data acknowledgments

ESRI thanks the organizations listed below for providing data used in the exercises.

The City of Manhattan, Kansas, and Riley County, Kansas, provided Manhattan, Kansas, data (chapters 2, 3, 4, 5, 13).

The U.S. Census Bureau provided the Washington, D.C., streets and landmarks data (chapters 6, 10, 11) and the attributes for the U.S. counties (chapters 7, 8).

The U.S. Geological Survey, EROS Data Center provided the Grand Canyon elevation data (chapter 12).

The David Rumsey Map Collection provided the 1887 historical map of Manhattan, Kansas (chapter 13). The David Rumsey Map Collection (www.davidrumsey.com) is one of the largest private map collections in the United States.

The Wilson County Mapping Department provided the Wilson County, North Carolina, data (chapters 14, 15, 16).

The U.S. Environmental Protection Agency provided the U.S. toxic sites data (chapters 17, 18, 19).

The U.S. Forest Service, Tongass National Forest, Ketchikan Area provided the Tongass National Forest, Alaska, forest stand data (chapter 20).

CONTENTS

Why learn how to program ESRI® ArcObjects™?

Think of ArcGIS® as a house that someone else has built and furnished for you. You move in and find that it has lots of rooms, great furniture, all the conveniences, artwork hanging on the walls, a swimming pool out back, everything you could ever want . . . except your own personal touch.

Learning how to program ArcObjects with VBA means learning how to make ArcMap™ and ArcCatalog™ your own. You may not even want to change them too much. Maybe remove a tool that you never use from a toolbar. Or take a certain menu choice and put it on a button. Or reduce some repetitive task to a single button click.

On the other hand, you might have plans to remodel the whole house.

Whatever the scope of your ambition, getting to know ArcObjects—and the concepts behind programming them—will give you a new level of confidence in your relationship with the software.

If you have little or no programming experience, begin with the book's first section, "Understanding VBA." There you'll learn about objects, properties, and methods; how to set variables; how to write procedures and link them to buttons on the user interface; how to use simple objects like message boxes; and how to employ programming constructs like If Then statements and Do While loops.

If you are already comfortable with an object-oriented language, especially Visual Basic® for Applications (VBA) or Microsoft® Visual Basic, you may want to start with the second section, "Understanding ArcObjects." There you'll learn about the ArcObjects architecture, how to read object model diagrams, and how to use programming interfaces. (If you have a good idea what the last sentence is talking about, you probably know enough to skip the first section.)

The third section of the book, "Using ArcObjects," leads you through a number of GIS programming tasks, such as adding layers to maps, symbolizing and classifying data, making feature selections, querying data, and editing feature attributes.

This is a workbook of hands-on exercises, supported by introductory discussions, diagrams, and screen captures. The exercises proceed step-by-step, showing every line of code that needs to be written and explaining its purpose. You will type lots and lots of code in this book. Please be assured that all the exercise code has been rigorously tested to work as instructed. Exercise code can fail, however, with even the smallest of typing mistakes. (But every mistake is an opportunity to learn.) If exercise code does fail, you can go through it yourself to check for typos. At different stages in the book, you will be introduced to many of VBA's error identification techniques.

You are encouraged to save your work along the way, but you don't have to because each exercise is a new starting point, already prepared with maps, data, and any VBA code you need. This is true even when the exercises build, as they often do, on work you have done previously. Every exercise also includes a map document with the correct results, in case you get stuck. Each chapter has from one to four exercises (except for chapter 1, which has none). The exercise length varies, but as a rule, chapters should take two to four hours to complete.

ArcGIS Desktop 8 was built with ArcObjects. These objects were designed and created following the COM (component object model) style of programming with C++. One of the rules of COM programming is that the signatures of an object's properties and methods should never change. This book was first written for ArcGIS Desktop 8.3 and all the code and examples still work using ArcGIS Desktop 9.1 and 9.2 without alterations. Updates to the book are mainly minor text edits and screen captures to accommodate changes to the ArcMap user interface.

To use this book, you must have a copy of ArcGIS Desktop 9.1 (or a later version) running on your computer. (You can use the ArcView®, ArcEditor™, or ArcInfo™ license level.) The CD that comes with the book contains the exercise data you need, plus additional resources, but does not include the software itself. For instructions on installing the data, refer to appendix B.

To resolve problems with the exercise data or to report mistakes in the book, send an e-mail to ESRI workbook support at *learngis@esri.com*. For the latest information, Q & A's, addenda, and errata, visit the book's companion Web site at *www.esri.com/ GTKArcObjects*.

CHAPTER 1

Programming with objects

In this book, you will learn how to program ArcObjects with Visual Basic for Applications (VBA). This raises two questions: what are ArcObjects and what is Visual Basic for Applications? You could say that ArcObjects is a set of programmable objects and Visual Basic for Applications is an object-oriented programming language. That's true, but not very helpful unless you know what object-oriented programming is.

Object-oriented programming

Object-oriented programming is a structure or design for computer programming languages. It resembles our own experience of the world. We see the world as divided into objects that have qualities, or properties, that behave in certain ways. For instance, a tree is an object. It has properties like a type (it could be an oak, a eucalyptus, or a palm), a height, an age, and lots of other things you could specify (evergreen—yes or no? fruit-bearing—yes or no? and so on). A tree also has behaviors. It may rustle in the wind or shed leaves in the fall. It grows from a little seed. One day it dies.

Similarly, in an object-oriented programming language, you work with objects that have properties and behaviors (which are called methods). Lots of these objects are familiar elements of software applications: buttons, tools, windows, and dialog boxes. Some are complex and specialized. Depending on your programming job, you might use computer objects that represent heart valves, honey bee hives, or hurricanes.

Computer programs solve problems or accomplish tasks: they beat you at chess, they guide rockets to Mars, they simulate a wide variety of human and natural processes. In object-oriented programming, you accomplish these tasks by giving instructions to objects that make them carry out their methods.

The syntax of these instructions varies somewhat from language to language, but the basic form is "Object.Method." (The period is pronounced "dot.") First comes the name of the object, then comes the behavior that you want from it. For example, to add a record to a table object, you would write something like:

```
Table.AddRecord
```

To make a tree object drop its leaves, you would write something like:

```
Tree.DropLeaves
```

Some methods take along extra information, called parameters or arguments. For example, in an instruction like:

```
Tree.Grow(10)
```

the argument tells the tree how much to grow—ten units of some kind or other.

What makes object-oriented programming interesting and rich is that objects have relationships to one another and effects on one another. In the real world, a tree has branches and branches have apples on them. These relationships are mirrored by computer objects. For example, you might be able to write an instruction to create a tree object (Tree.Create) but not to create a branch object. Branch.Create doesn't work because branches don't exist independently of trees—and this is defined by the properties and methods of your objects. Instead, to make a branch, you first create a tree and then run its GrowBranch method.

```
Tree.Create
Tree.GrowBranch
```

If you want to make an apple, it might be Tree.Create, followed by Tree.GrowBranch, followed by Branch.GrowApple.

Suppose you want to be elaborate and make a boy object eat an apple object. You would first write a series of instructions, like those above, to make an apple, and then another series to make a boy. Having produced these two objects, you would write an instruction like:

```
Boy.Eat(AnApple)
```

When the boy eats the apple, consequences ensue. For example, the apple will no longer exist and the boy's IsHungry property might change from Yes to No.

You can see that in object-oriented programming, it's important to learn the relationships among the objects you work with: which ones you can make, which ones make others, which ones have which properties and methods, and which ones are affected by the properties and methods of others. The need to understand these relationships—which can be very complicated—has led to the creation of a special system for diagramming them. You will start seeing these diagrams in the second part of the book. They look like this:

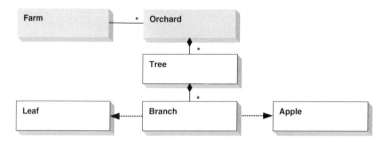

Once you know what all the arrows, boxes, diamonds, lines, and stars mean, you'll be able to tell from a diagram that a branch object can make an apple object; that apple and leaf objects can exist independently of tree and branch objects; and that if you want a tree to do something, you first have to specify which orchard the tree is in and which farm the orchard is on. In other words, diagrams give you the rules you need to follow when you write code.

When you get to the second part of this book, it will be helpful to have a look at some of these diagrams, including the ones called ArcMap, Geodatabase, Display, Geometry, and Styles. All the ArcObjects diagrams are included as PDF files on the CD that comes with this book.

Visual Basic for Applications (VBA)

Visual Basic for Applications, a simplified version of Visual Basic, is one of many object-oriented programming languages. You can use other languages to program ArcObjects (like C++ or Visual Basic), but VBA comes included with ArcGIS Desktop. If ArcMap and ArcCatalog are loaded on your computer, so is VBA.

The main difference between VBA and other object-oriented programming languages is that VBA was designed to be embedded within applications. Programmers working in languages like C++ typically build software applications from scratch with a set of objects, while VBA programmers customize applications, like ArcMap, that come with VBA inside them. VBA programmers can use as much of the application's built-in functionality as they want.

VBA has its own set of development tools, many of which you will use in this book. For example, it has windows for organizing and storing the code you write; tools for creating dialog boxes and their components (buttons, drop-down lists, scrolling boxes, and so on); and tools for debugging code.

Code is organized in procedures. A procedure includes all the instructions needed to accomplish some clearly defined task, like printing a map or making a boy eat an apple. Procedures can be linked to each other so that when one finishes it tells another to begin. For example, when a procedure that spell checks a document is finished, it might "call" another procedure that prints the document.

You may sometimes have a chain of procedures calling one another, but the first link in the chain requires a human touch. Actions like opening an application, clicking a button, or moving the mouse pointer are called "events" and events are what make procedures run. The example below shows the procedure that runs when a person clicks the Full Extent button in ArcMap.

User clicks button   Code runs

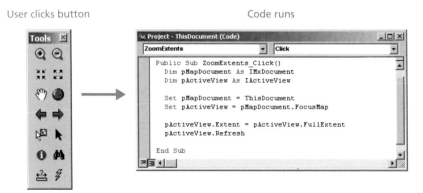

ArcObjects

ArcObjects are a set of computer objects specifically designed for programming with ArcGIS Desktop applications. ArcObjects include things like data frames, layers, features, tables, cartographic symbols, and the pieces that make up these things: points, lines, polygons, records, fields, colors, and so on.

As an ArcGIS user, you have been using ArcObjects all along. Click the Save button in ArcMap and a procedure runs to save your map document. Click the Draw Point tool on a map and a procedure runs to draw a point graphic where you clicked. The map document, the Save button, the Draw Point tool, and the point graphic are all ArcObjects. In fact, ArcMap and ArcCatalog are both built from ArcObjects.

In this book, you will use ArcObjects to customize and enhance ArcMap (and also just to get underneath the hood and see how ArcMap works). ArcObjects can be used to program other applications as well. Thanks to a design standard called COM, which provides guidelines to the programmers who develop objects, you can use ArcObjects to put map functionality in applications like Microsoft Word or Microsoft Excel (and you can also put word-processing or spreadsheet functionality in ArcMap).

One thing you won't do in this book is master the universe of ArcObjects—there are far too many of them. You will get to know a lot of the common ones, however. More importantly, you'll get to know the principles of working with them and you'll learn how to read object model diagrams, so that when you're done with the book, you'll be ready to explore ArcObjects on your own.

Building a custom application

Organizing commands on a toolbar
Making your own commands
Storing values with variables

ArcMap and ArcCatalog have plenty of toolbars, menus, and commands (buttons and tools), enough and more for most users' needs. However, they will never have everything that everyone needs, and will always have something that some people simply do not. Fortunately, with a little training you will be able to change things around, move toolbars, menus, and commands, delete ones you don't use, and create the ones that no one could ever have predicted you'd need.

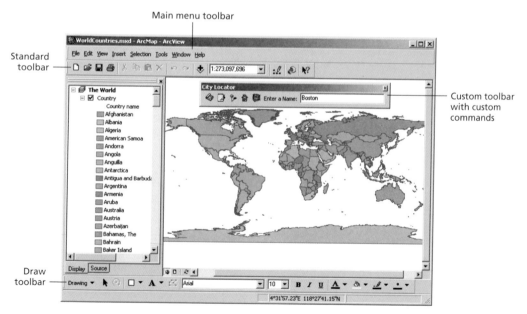

Some of your ArcObjects work isn't programming at all, but customizing the user interface. Objects, like toolbars, buttons, and tools, can be created and rearranged without any programming. Of course, if you create a new button and want it to do something, then you have to write the code that runs when the user clicks it.

If you've used any software package long enough, there are probably some repetitive actions you do on a daily or weekly basis. Maybe you make a lot of feature selections in ArcMap and you'd like to be able to do that by clicking buttons instead of choosing items from the Selection menu. So you make a toolbar and put all the selection choices on it. You don't do any programming, because the Selection buttons already have code.

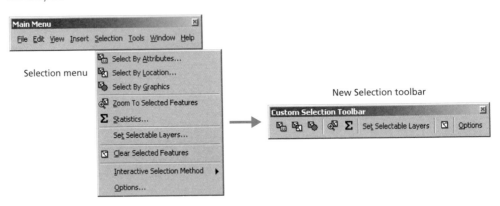

When you need a completely new function, you can create your own buttons and tools and put them on toolbars. And then you get to write code for them.

Your new button

Organizing commands on a toolbar

The graphic user interface (GUI) for ArcMap and ArcCatalog is made up of toolbars, menus, and commands. Toolbars contain commands or menus. Commands come as either buttons or tools. Menus provide pull-down lists of commands or of other menus.

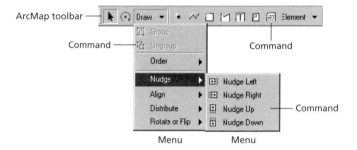

ArcMap toolbar

Command

Command

Command

Menu   Menu

To make changes to the user interface, you use the Customize dialog box, shown below. You use the Toolbars tab to create, delete, rename, and reset toolbars, and to turn them on and off. The Commands tab displays commands and menus, which you can drag from the Customize dialog box onto toolbars. The Options tab contains security and other settings.

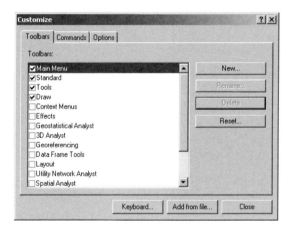

Exercise 2a

You are a GIS programmer for the planning department of Manhattan, Kansas. When people come to the planning department office for parcel maps, a planning department staff member uses ArcMap to locate the parcels and print the maps.

To free up staff members' time, you have been asked to create a parcel viewer application that anyone can use.

Users of the parcel viewer application will need to find parcels, label them with identification numbers and values, zoom in on them, pan to center them, and print a map. They'll also need to zoom back out to see the entire city. In ArcMap, you'll create a toolbar with commands for only those actions.

1 Start ArcMap and open **ex02a.mxd** in the **C:\ArcObjects\Chapter02** folder. C:\ArcObjects is the assumed installation folder for this book's exercise and sample data. If you installed your data in a different folder or drive, go there now, locate the Chapter02 folder, and open the map.

The map contains layers for the city boundary, parcels, and streets.

2 Click the Tools menu and click Customize. Click the Toolbars tab, unless it's already active.

On the Toolbars tab, checked toolbars are visible, unchecked are not. Your list of toolbars will vary according to which ArcGIS extensions you have loaded.

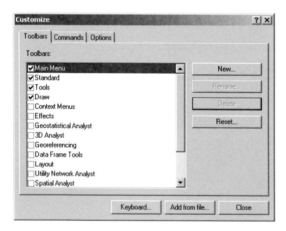

3 Click New.

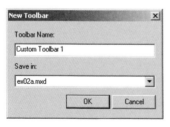

The New Toolbar dialog box opens.

4 Replace the toolbar name with **Parcel Viewer**. Make sure ex02a.mxd is selected under Save in.

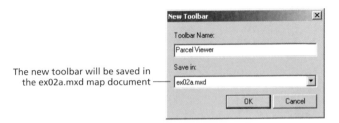

The new toolbar will be saved in the ex02a.mxd map document ——

5 Click OK.

Parcel Viewer is added to the list of toolbars in the Customize dialog box. The small, empty undocked toolbar will expand as you add commands to it.

You'll now move the Find command from the Tools toolbar to the Parcel Viewer toolbar. People will use the Find dialog box to locate a parcel by typing a parcel ID number.

6 With the Customize dialog box open, drag Find from the Tools toolbar to the Parcel Viewer toolbar. (If your Tools toolbar is not displayed, you can go to the Toolbars tab of the Customize dialog box and check its box.)

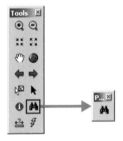 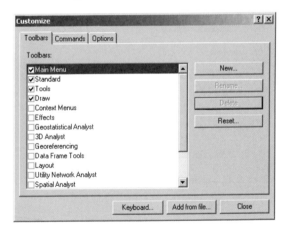

Find is no longer on the Tools toolbar, because moving a command from one toolbar to another doesn't make a copy of it. In the following steps, you will learn how to put copies of a command on different toolbars. That way Find can be on both the Tools toolbar and the Parcel Viewer toolbar.

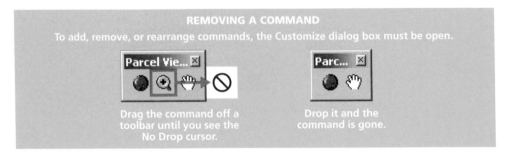

REMOVING A COMMAND
To add, remove, or rearrange commands, the Customize dialog box must be open.

Drag the command off a toolbar until you see the No Drop cursor.

Drop it and the command is gone.

7 In the Customize dialog box, click the Commands tab.

The Commands tab displays two lists, Categories and Commands. Categories organizes commands by their functionality. Your list of categories may vary according to which ArcGIS extensions you have loaded.

The Commands list contains commands that you drag onto toolbars. Unlike dragging from one toolbar to another, commands remain in the Commands list. So you can drag as many copies to as many toolbars as you need.

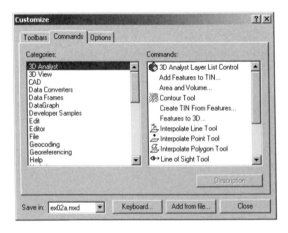

In the following steps, you will drag commands to the Parcel Viewer toolbar to zoom in on the map, pan around it, print it, and zoom out to the full extent of the city.

CHAPTER 2 • BUILDING A CUSTOM APPLICATION

8 In the Categories list, scroll down and click Pan/Zoom.

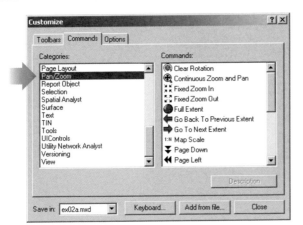

9 From the Commands list, drag Full Extent to the Parcel Viewer toolbar.

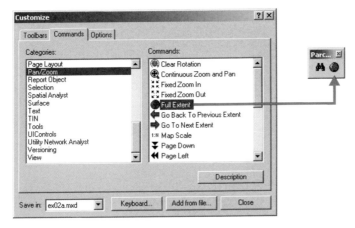

10 Add the Zoom In and Pan commands to match the graphic below.

Next, you will add the Data View and Layout View commands to the Parcel Viewer toolbar. This way users can go back and forth between looking at the data and seeing how the map will look on paper with all its cartographic components like north arrow, scale, and title.

11 In the Categories list, click View. Drag the Data View command from the Commands list onto the Parcel Viewer toolbar. Do the same for the Layout View command. Your toolbar should look like the one shown below.

Since some people may not be familiar with the ArcMap icons, you will use text instead of icons for the Data and Layout View commands.

12 On the Parcel Viewer toolbar, right-click Data View and click Text Only.

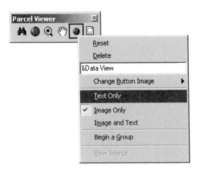

The icon is gone and only text displays. To keep all the commands on the same line, you may have to widen the toolbar.

13 Right-click Layout View and click Text Only.

The two commands are now displayed on the toolbar with their names. To make them look more like buttons, you will add separator lines.

14 On the Parcel Viewer toolbar, right-click Data View and click Begin a Group. Do the same for Layout View.

Your toolbar should now look like this:

15 In the Categories list, click File. From the Commands list, drag Print to the Parcel Viewer toolbar.

16 Right-click Print and click Begin a Group.

The Parcel Viewer toolbar now contains the commands you need to find a parcel and print a map. All other toolbars can be turned off.

17 In the Customize dialog box, click the Toolbars tab. Uncheck all toolbars except Main Menu and Parcel Viewer.

Although you don't need it, ArcMap requires the Main Menu toolbar to be turned on.

18 In the Customize dialog box, click Close.

19 Dock the Parcel Viewer toolbar below the Main Menu.

Your ArcMap application should look like the graphic below.

20 Test the Parcel Viewer toolbar by following the instructions below.

— On the Parcel Viewer toolbar, click Find.

— If needed, click the Features tab.

— In the Find drop-down area, type the parcel ID, **71400**.

— Click the Find button. The parcel record appears in the object found list.

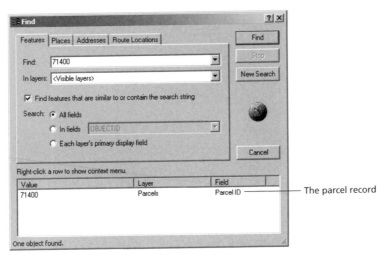

— In the object found list, right-click the parcel and click Flash feature.

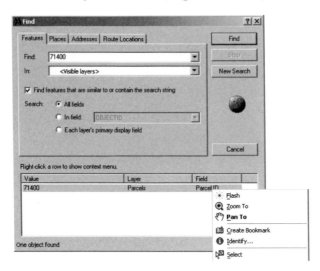

— If needed, move the Find window so you see the parcel flash in the display area.

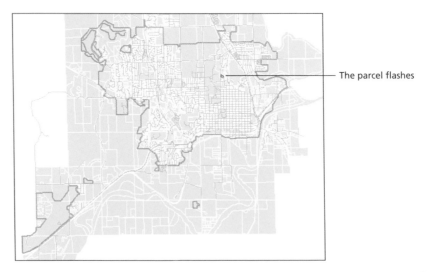

The parcel flashes

— In the dialog, right-click the parcel row again and click Zoom to feature(s).

— Right-click the parcel again and click Select feature(s).

— Close the Find window.

— On the Parcel Viewer toolbar, click Layout View to see what the map looks like.

The parcel map displays, ready to print.

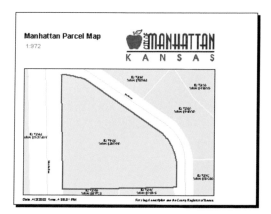

During the exercise, you turned off all toolbars except Main Menu and Parcel Viewer. You can turn them back on at any time, but you won't need them in the following exercises.

You can use a shortcut to open the Customize dialog box. Double-click on a blank area in the toolbar area.

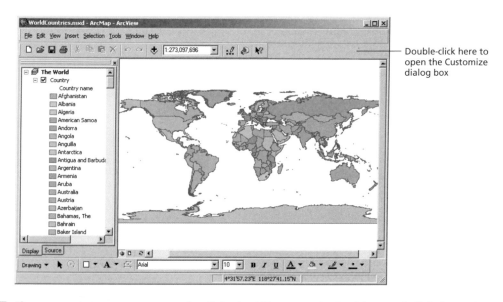

Double-click here to open the Customize dialog box

21 If you want to save your work, click the File menu in ArcMap and click Save As. Navigate to **C:\ArcObjects\Chapter02**. Rename the file **my_ex02a.mxd** and click Save. If you are continuing with the next exercise, leave ArcMap open. Otherwise close it.

Making your own commands

In the previous exercise, you made a toolbar and added some preprogrammed ArcMap commands to it. When your application has to perform tasks that aren't already part of ArcMap, you make your own commands and write code to make them work.

You use the Customize dialog box to create commands called user interface controls, or UIControls for short. UIButton, UITool, UIComboBox, and UIEditBox are the four types of UIControl you can make.

UIButton UITool   UIComboBox   UIEditBox

As with any button, clicking a UIButton makes something happen. For example, when you click the Save button, the map is saved. UITools work by clicking them and then clicking in the map display area. Only one can be active at a time. After you've clicked on a tool, its cursor symbol appears when you move into the map display area. For example, when you click the Pan tool, the cursor you move over the map looks like a hand. You click and drag on the map to pan to different locations.

Map display area

Tool's cursor symbol

To get a UIControl to do something, you write code with Visual Basic Editor. There, VBA organizes your code into projects, code modules, and procedures.

Projects   Code module   Procedure   Visual Basic Editor

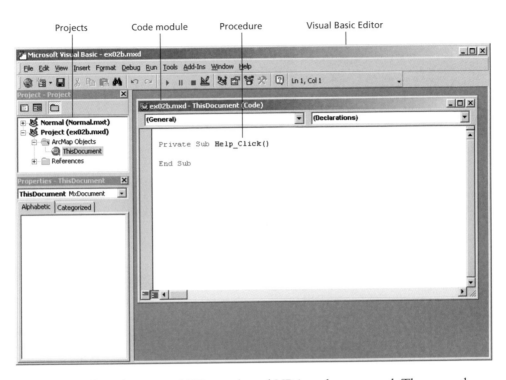

Projects are files where your UIControls and VBA code are stored. There are three types of projects: map documents (.mxd files), base templates (.mxt files), and the normal template (the Normal.mxt file stored in your personal profiles folder).

Say you have a map document called rivers.mxd and it contains rivers and other water-related layers. And you need to make a UIButton to calculate water flow rates. You would put the Flow Rate button and its code in the rivers.mxd project. It wouldn't make sense to put the button into other map documents, since its code will be specific to calculate flow on rivers.

Say you need to make a UITool that reports any polygon's area in square feet, acres, and hectares. This tool is not specific to any project. In fact you want it available whenever you have polygon layers displayed in any map document. You would put this Area Reporter tool and its code in the Normal.mxt project. Every time ArcMap starts up, it reads the normal template to see if it contains any UIControls and code that needs to be available at all times. If your organization uses base templates, you can also store UIControls and code in them.

Projects organize code with code modules, windows that you type into. You make as many code modules as you want or need. However, every project has a code module called ThisDocument. The ThisDocument code module is the place for you to write code for any UIControls that you make in a particular project.

Within each code module, your VBA code is organized into procedures, lists of VBA instructions that perform a task. A procedure to cut the lawn might look like this:

```
Start CutTheLawn
    Get lawn mower from garage, fill with fuel if needed
    Remove sticks and other debris from lawn
    Start lawn mower and cut grass
    Turn mower off, clean, and return it to garage
    Drink cool beverage
End CutTheLawn
```

VBA has four types of procedures: event, subroutine, function, and property. Event procedures correspond to user actions. For example, a UIButton has an event procedure called Click. You code the click event procedure so it will run when the user clicks the button. The other three procedures are all called into action by event procedures. So no code runs unless the user does something (causes an event).

All procedures have a first and last line, which are called wrapper lines. Your code goes between them. The first line contains either the Public or Private keyword. Private procedures can only be called on by procedures in the same code module. Public procedures can be called on by procedures in other code modules. The first line also contains a keyword identifying the type of procedure (Sub, Function, Property), the name of the procedure (for instance, cmdQuit_Click), and parentheses. (VBA identifies both subroutine and event procedures with the Sub keyword.) The parentheses are required. Here they're empty, but in later exercises, you'll learn what can go into them.

```
Public Sub cmdQuit_Click ()
    MsgBox "ArcMap is about to close"
End Sub
```

The last line in a procedure contains the End keyword and the keyword identifying the type of procedure. The click event procedure above has one indented line of code. Indenting makes your code easier to read without slowing it down.

MsgBox is a Visual Basic function that displays the quoted text message in a dialog box. When someone clicks the Quit button, the MsgBox line of code is executed, and the dialog box below appears. Visual Basic comes with many prewritten functions like MsgBox. You'll learn how to use it and other functions, and how to write your own, too.

Exercise 2b

People are now using the Parcel Viewer application to print their parcel maps. Occasionally, some people need help finding their parcel and making a map. You will create a button to tell people who to contact if they need help.

In this exercise, you will make a button. You will write code in the button's click event, so someone can click it to get assistance with their map.

1 Start ArcMap and open **ex02b.mxd** in the **C:\ArcObjects\Chapter02** folder.

When the map opens, you see the parcels of Manhattan and the Parcel Viewer toolbar.

2 In ArcMap, click the Tools menu and click Customize. Click the Commands tab.

3 In the Categories list, scroll all the way down to the bottom of the list and click [UIControls]. Make sure ex02b.mxd is selected under Save in.

The Commands list is empty. As you make UIControls, they will be listed there.

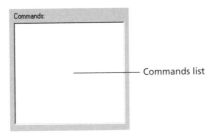

Commands list

4 Click the New UIControl button.

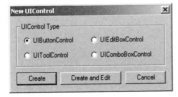

5 With the radio button next to UIButtonControl selected, click Create.

You see the new button, Project.UIButtonControl1, in the Commands list. Its name has the "Project" prefix, because you selected ex02b in the Save in combo box. Had you selected Normal.mxt for Save in, the button's name would have "Normal" as a prefix. UIControls saved in a base template have a "TemplateProject" prefix. Since you saved the button in the ex02b.mxd file, you and your users must open ex02b.mxd to see and use this new button. The Normal.mxt file is a file that ArcMap reads every time it starts up. Any UIControls you save in Normal.mxt always appear, regardless of the .mxd file you open.

6 In the Commands list, click Project.UIButtonControl1 and change the name to **Project.Help**.

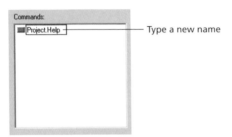 Type a new name

7 Drag Project.Help to the Parcel Viewer toolbar.

 Add the button here

You will change the button's icon to display text only.

8 Right-click the new Help button and click Text Only.

9 Right-click the Help button and click Begin a Group to add a separator line.

Now you will write some VBA code to make the new Help button work.

10 Right-click the Help button and click View Source.

The Visual Basic Editor window opens. It contains other windows, like Project, Properties, and a code window called ThisDocument. Depending on your VBA settings, these windows may be different sizes or in different positions compared to the ones in the following graphic.

The Project window below, sometimes called the Project Explorer, shows two projects: Normal (Normal.mxt) and Project (ex02b.mxd). Although the Normal project is always present, you will add UIControls and write code in the current project that corresponds to the currently opened map document, ex02b.mxd.

Event procedures for the project's UIControls are stored in a code window called ThisDocument. That's why when you right-click on a UIControl and click View Source, the ThisDocument code module opens. If it is a new UIButton, wrapper lines for the button's click event procedure are automatically added.

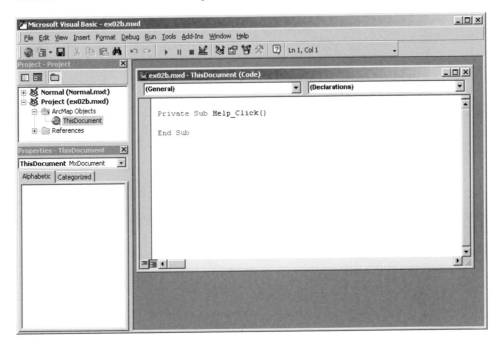

11 Between the wrapper lines, indent for a new line of code. Type **msgbox**, press the space bar, and stop typing for a moment.

```
msgbox
```

A yellow help tip pops up to display help for using the MsgBox. The words and phrases separated by commas are called arguments. To make unique-looking message boxes, you fill in different information for each argument.

The word "Prompt" is bold because it is the first argument you should type. For Prompt, you type a text string to be displayed in the message box's dialog box. Arguments in square brackets are optional. For a description of each argument, look up MsgBox in the online help.

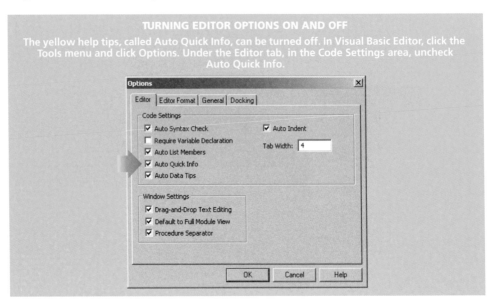

TURNING EDITOR OPTIONS ON AND OFF

The yellow help tips, called Auto Quick Info, can be turned off. In Visual Basic Editor, click the Tools menu and click Options. Under the Editor tab, in the Code Settings area, uncheck Auto Quick Info.

12 Finish typing the MsgBox line of code by adding the following text string, quotes and all.

```
msgbox "For help dial extension 25"
```

The click event procedure is now ready to test.

13 Highlight the procedure as shown below.

VBA checks your syntax and, when it's okay, converts it to uppercase for you as needed. You typed **msgbox** and VBA changed it to **MsgBox**. It's a good idea to do your typing in lowercase and let VBA check your syntax and handle capitalization.

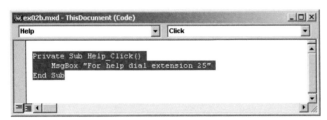

14 On the Standard toolbar, click the Run button (or press F5).

The Visual Basic Editor Standard toolbar has a button that looks like a VCR's Run button. Clicking that button tells VBA to start executing the lines of code in the selected procedure.

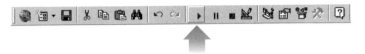

The line of code runs, ArcMap comes to the foreground, and the help message appears.

Highlighting a procedure and clicking the Run button is one way to test a procedure without leaving Visual Basic Editor. Parcel Viewer users, however, will not use the Run button to run this procedure. So next you'll use the Help button just like a user would.

15 Click OK on the message box.

16 Close Visual Basic Editor.

17 On the Parcel Viewer toolbar, click Help.

The message displays.

18 Click OK.

In this exercise you made a UIButton, put it on a toolbar, and wrote VBA code in its click event procedure.

19 If you want to save your work, click the File menu in ArcMap and click Save As. Navigate to **C:\ArcObjects\Chapter02**. Rename the file **my_ex02b.mxd** and click Save. If you are continuing with the next exercise, leave ArcMap open. Otherwise close it.

Storing values with variables

In the previous exercise you made a button and named it Project.Help. The button's name is a value that gets stored in and saved with the map document. Values like a button's will probably never change. Some values, however, do change. Say you want to keep track of the temperature outside your home. As it gets warmer or colder outside, the temperature value changes. You use variables to store values that might change over time.

To make a variable you write a line of code. In the line, you make up a variable name and use the equals sign to assign it a value. The line below makes a variable named Temperature and sets it equal to 32.

```
Temperature = 32
```

This line of code is called an assignment statement because you assigned a number to the variable using an equals sign. After the code runs, Temperature contains or represents the number 32. In any subsequent lines of code, if you need the temperature value, you use the Temperature variable. For example, you could display the value in a message box with the following line of code:

```
MsgBox Temperature
```

The value inside the Temperature variable displays here

As the day goes on, the number stored for Temperature will vary depending on the weather. Because the value will vary, Temperature is called a variable.

The Temperature variable currently contains 32. A variable's contents can change by the mere writing of another assignment statement. As it gets warmer outside, a new number is assigned. When the line of code below runs, the old number is cleared out and the new number is assigned.

```
Temperature = 50
```

The Temperature variable now contains 50.

A variable can be assigned a value that is the result of a math equation. Equations are arranged to the right of the equals sign using operators like +, −, *, and /. The equation below subtracts 5 from the current temperature. The resulting value of 45 is assigned to the variable.

```
Temperature = 50 - 5
```

The temperature dropped five degrees, the new value is assigned, and Temperature now contains 45.

Variables can be substituted for numbers. For example, you could have subtracted 5 from the current temperature with the following line of code. The math equation, Temperature – 5, is solved before the variable's value is assigned.

```
Temperature = Temperature - 5
```

All math is performed from left to right, except for operations in parentheses, which are evaluated first. In complex equations you use parentheses to control the order of evaluation.

In places where it gets really cold (below 50 degrees Fahrenheit) and windy (wind speeds above 5 miles per hour), temperature gets measured in wind chill factor, a measure of how fast your body loses heat. With a temperature of 35° F and a wind speed of 20 mph, the temperature feels like 24° F to your body, but water will not freeze.

The equation below computes the wind chill factor given temperature (T) and wind velocity (V); the result is assigned to the Chill variable. The operations inside the three sets of parentheses below are evaluated first, and the results are added to 35.74.

```
Chill = 35.74 + (0.621 * T) + (35.75 * V) + (0.4275 * T * V)
```

Besides numbers, variables can contain other values, like character strings (text). In an assignment statement to create a string variable, the characters must be quoted. The colon and space after *is* have been added so that a number can go there.

```
Message = "The current temperature is: "
```

To add the temperature next to the message, you use the ampersand character (&), which concatenates the two values.

```
MsgBox Message & Temperature
```

When this line of code runs, the message and temperature display as one value.

To keep the temperature updated, someone must enter new numbers. A dialog box, called InputBox, has an area to type in values.

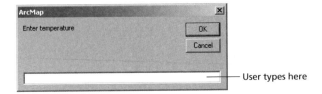

User types here

An InputBox can be used in an assignment statement. The value typed in the input area is assigned to the variable. When you set variable values with an InputBox, its arguments, like the text message below, go in parentheses.

```
Temperature = InputBox ("Enter temperature")
```

When the user types 85 and clicks OK, the Temperature variable contains 85.

Unless otherwise instructed, VBA sets aside 16 bytes of space to store a variable's value. With that much space, you can store the number 1.797693134862315 E308. E308 means that the decimal place in 1.797693134862315 can be moved 308 places to the right. That's like taking away the decimal place and adding 293 zeros to the end of that number. If the variable only needs to store a temperature, that's a lot of wasted space.

You cut down a variable's storage space to size by telling VBA what values the variable should accommodate. This is called declaring or dimensioning a variable. You tell VBA the variable name and its data type, which could be number, string, date, or any of several other data types.

Declaring a variable takes one line of code: here it's the Dim keyword (for dimension), Temperature (the variable name), As (a keyword), and Integer (the data type).

```
Dim Temperature As Integer
```

Dim is only one of several keywords used to declare variables. You'll learn about others—public, private, and static—in later chapters.

Now that Temperature is declared as an Integer data type, it will only require 2 bytes of storage space, a savings of 14 bytes. You can learn more about VBA's data types and their storage requirements in the online help. For a list of sizes and value ranges, search for Data Type Summary, as shown in the following graphic.

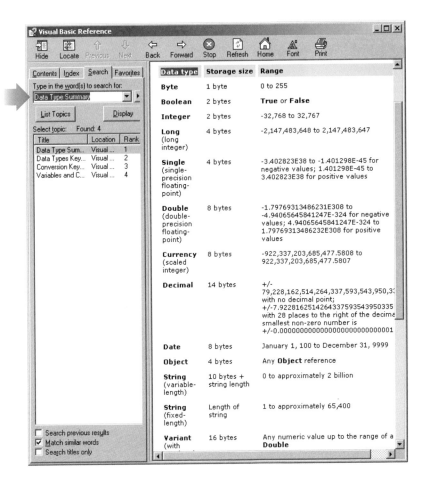

The following data type table appears in the Visual Basic Reference window:

Data type	Storage size	Range
Byte	1 byte	0 to 255
Boolean	2 bytes	**True** or **False**
Integer	2 bytes	-32,768 to 32,767
Long (long integer)	4 bytes	-2,147,483,648 to 2,147,483,647
Single (single-precision floating-point)	4 bytes	-3.402823E38 to -1.401298E-45 for negative values; 1.401298E-45 to 3.402823E38 for positive values
Double (double-precision floating-point)	8 bytes	-1.79769313486231E308 to -4.94065645841247E-324 for negative values; 4.94065645841247E-324 to 1.79769313486232E308 for positive values
Currency (scaled integer)	8 bytes	-922,337,203,685,477.5808 to 922,337,203,685,477.5807
Decimal	14 bytes	+/- 79,228,162,514,264,337,593,543,950,3? with no decimal point; +/-7.9228162514264337593543950335 with 28 places to the right of the decima smallest non-zero number is +/-0.0000000000000000000000000001
Date	8 bytes	January 1, 100 to December 31, 9999
Object	4 bytes	Any **Object** reference
String (variable-length)	10 bytes + string length	0 to approximately 2 billion
String (fixed-length)	Length of string	1 to approximately 65,400
Variant (with	16 bytes	Any numeric value up to the range of a **Double**

Exercise 2c

People are using the Parcel Viewer application to locate parcels and print their maps but using pocket calculators to compute the tax for a parcel. They'd like the application to compute the taxes for them. In this exercise, you'll create a UIButton that calculates a parcel's tax. You'll make an input box that asks the user for the parcel's value and then displays a message box with the tax.

1 Start ArcMap and open **ex02c.mxd** in the **C:\ArcObjects\Chapter02** folder.

When the map opens, you see the parcels of Manhattan and the Parcel Viewer toolbar.

2 In ArcMap, click the Tools menu and click Customize. Click the Commands tab.

3 In the Categories list, scroll down and click UIControls. Make sure ex02c.mxd is selected under Save in.

In the Commands list, you see the Help button you created in the previous exercise.

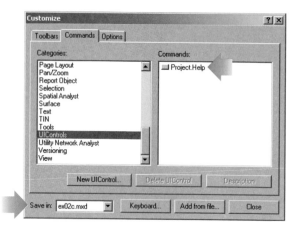

4 Click the New UIControl button.

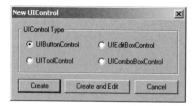

5 With the radio button next to UIButtonControl selected, click Create.

In the Commands list, you see a new button named Project.UIButtonControl1.

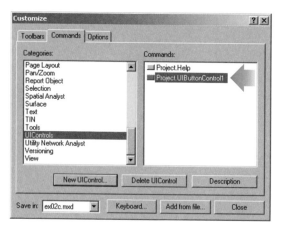

6 In the Commands list, click Project.UIButtonControl1 and change the name to **Project.CalculateTax**.

7 Drag Project.CalculateTax to the Parcel Viewer toolbar and drop it to the right of the Pan tool.

You will change the button's icon.

8 Right-click the new button, point to Change Button Image, and click Browse.

9 In the Open file browser, navigate to **C:\ArcObjects\Data** and click **Dollar.bmp**. Click Open.

The dollar sign appears.

Now you will write code to make the Calculate Tax button work.

10 Right-click the Calculate Tax button and click View Source.

The Visual Basic Editor window opens. You see the empty wrapper lines for the Calculate Tax click event procedure in the ThisDocument code window. You also see the Help click event procedure you coded in the previous exercise.

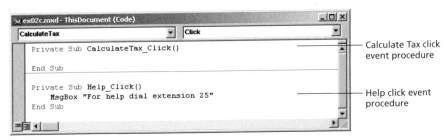

11 Between the wrapper lines, indent and declare the following variables.

```
Dim curParcelValue As Currency
Dim curTaxValue As Currency
Dim datToday As Date
```

The Currency data type has two zeros to the right of a decimal to store dollars-and-cents values. The Date data type is capable of storing the current time, day, date, month, and year.

Variable names must begin with a character and not exceed 255 characters. They cannot contain spaces, periods, or special characters (%, *, &), or be the same as a VBA keyword like If, Then, Sub, And, Or.

Within the rules, you can name variables as you like. Some programmers add a prefix to variable names to indicate the variable's data type. For example, all string variables might be named with the prefix "str": strName, strAddress, strPhoneNumber. In the code above, *cur* is a prefix for Currency variables and *dat* for Date variables.

Next you will use an InputBox to set the parcel value variable. Since InputBox has three arguments—user message, window title, and default value—the line of code that creates it will be rather long. To make it easier to read, you'll use the line continuation character (an underscore) to put each argument on its own line.

12 Enter the following line of code, putting a space before each underscore.

```
curParcelValue = InputBox ( _
    "Enter a parcel value", _
    "Parcel Viewer", _
    100000)
```

Even though you see four lines, VBA recognizes them as one line of code and executes it as such. You can only use the line continuation character between arguments. You will get an error if you try to break a text string as shown below.

```
"Parcel _
Viewer",
```

13 Enter the following code to compute the tax value.

```
curTaxValue = (curParcelValue * 0.02) + 8.55 + 11
```

Residential parcels are taxed at 2 percent of their value, plus an $8.55 convention center fee, and $11.00 for the city's new fire truck.

The number in the input box is multiplied by 0.02, then 8.55 and 11 are added. The result of the calculation is assigned to curTaxValue.

14 Enter the following code to set the date variable.

```
datToday = Now
```

"Now" is a Visual Basic function. You'll learn how functions work in a few chapters. For the time being, you can think of "Now" as a predefined variable that contains the current date and time.

All the necessary information has been gathered and evaluated. What remains is to give the information to the user in a message box.

15 Enter the following code to display a message with the parcel's tax amount and today's date. Use the vbInformation constant for the type of message box.

```
MsgBox "The tax value is: $" & curTaxValue, _
    vbInformation, _
    datToday
```

Normally, after you type the first argument and a comma, you see a list of message box types to be used for the second argument, as shown below. You didn't see the list because you used line continuation.

```
MsgBox "The tax value is: $" & curTaxValue,
```

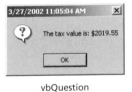

The choices in the list are called constants. Each has a vb prefix and produces a different dialog box. Examples of vbCritical, vbQuestion, and vbExclamation are shown below.

vbCritical

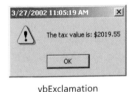

vbQuestion

vbExclamation

The third argument, datToday, is the message box's title. The datToday variable, created in the previous step, contains today's date. Without the third argument, the message box title will read either "ArcMap" or "ArcCatalog," depending on which application you wrote the procedure in. Your code should match that shown below.

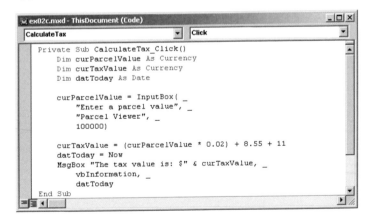

Before testing the code, you will add a line of code to help identify errors.

16 Move to the top of the code module and type the following line so that it is the first line of code. Press Enter to add an empty line between Option Explicit and the first procedure.

```
Option Explicit
```

The top of the code module is called the General Declarations area. In this area you can declare variables and set options like Option Explicit.

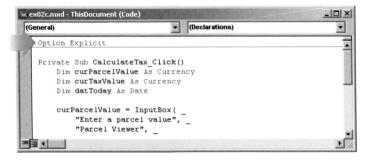

With Option Explicit on, your code will not run if you have any undeclared variables. This is VBA's way of automatically locating spelling mistakes. In the following code, one line declares the variable (Test), one sets it, and the other uses it.

On the third line, the variable has been intentionally misspelled Fest.

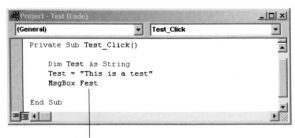

Fest should be *Test*

Without Option Explicit, the three lines of code above execute without error and result in the following empty message box. Fest is recognized as a variable, but because it isn't declared or set it contains no value.

With Option Explicit on, when an attempt is made to execute the code, Fest is identified as an undeclared variable, and the following error message displays:

In the code window, VBA highlights the misspelled variable.

With Option Explicit added to the code module, the Calculate Tax button is ready to test.

17 Close Visual Basic Editor.

18 Test the Calculate Tax button by following the instructions below.

— On the Parcel Viewer toolbar, click Calculate Tax ($).

— For the value, replace 100000 with **150000**. (If you enter text instead of a number, or if you click the Cancel button, an error message will appear. In chapter 5, you'll learn how to use branching statements to help avoid these errors.)

— Click OK.

— Click OK.

You have now created two UIButtons and coded their click event procedures. Users click the buttons and your code runs. Next you are going to code a second type of procedure called ToolTip.

Code in a ToolTip event procedure runs when a user moves their mouse over the top of a command. There is no clicking involved. Below, a user hovers the mouse over the Zoom In command and a yellow tooltip appears with the name of the command.

19 Right-click the Calculate Tax button and click View Source.

CalculateTax is selected in the object list, because you right-clicked it. You see CalculateTax's click event procedure and the code you already added. To add other event procedures for CalculateTax, you select them from the procedure list. You will add the ToolTip event procedure.

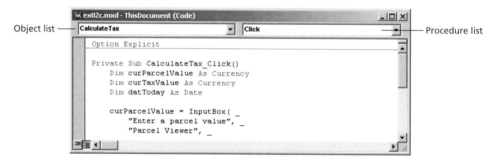

20 In the procedure list, click the drop-down arrow and click ToolTip.

The wrapper lines for the ToolTip event procedure are added.

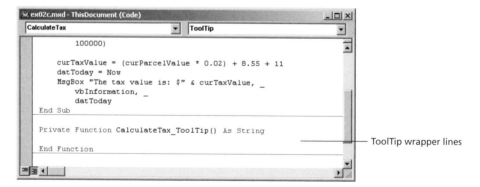

21 Add the following line of code to set the tooltip's help string.

```
CalculateTax_ToolTip = "Calculate Tax"
```

Code in a ToolTip event procedure will not run if Visual Basic Editor is open.

22 Close Visual Basic Editor.

23 Hover your mouse over the Calculate Tax button to see its tooltip.

24 If you want to save your work, click the File menu in ArcMap and click Save As. Navigate to **C:\ArcObjects\Chapter02**. Rename the file **my_ex02c.mxd** and click Save. If you are continuing with the next chapter, leave ArcMap open. Otherwise close it.

Creating a dialog box

Using controls to build a form

In the previous chapter, you called up message boxes and input boxes just by using MsgBox or InputBox in lines of code. You could do that because both of them are canned dialog boxes—precoded, preset, prefab. But because they're canned, they can only do so much. You can't add buttons, slider bars, or input boxes to them. To get a user's name, address, and phone number, you'd have to call three different input boxes.

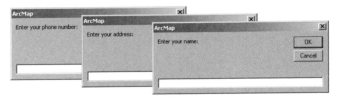

Fortunately, there's a better way to do it. You can make your own dialog boxes with as many input boxes, check boxes, and slider bars as you need.

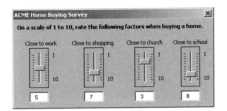

Using controls to build a form

No matter what they look like to a user, all these dialog boxes are called forms by programmers. The form below has five input boxes and three buttons. After filling out the form, the user clicks Add Customer to enter the data into a customer database.

 —— Form

In Visual Basic Editor, the blank canvas for making a user dialog box is also called a form. The objects on a form, like buttons, input boxes, and text, are called controls.

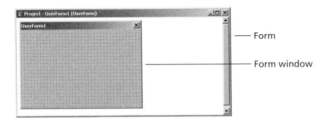 —— Form

—— Form window

In the same way that you drag commands from the Customize dialog box onto a toolbar, you drag controls onto a form from Visual Basic Editor's Toolbox.

Toolbox ——

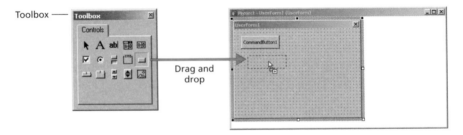

Drag and drop

Label controls are used to add descriptive information.

Label control ——

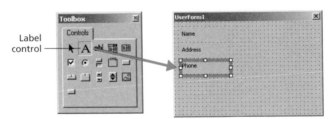

Text boxes provide an area for the user to type in information.

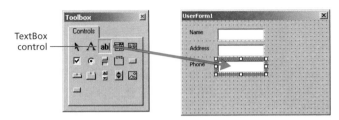

TextBox
control

Combo boxes provide a list of choices that users click a drop-down arrow to see.

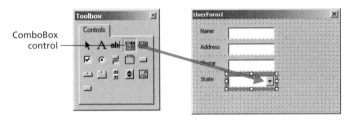

ComboBox
control

While designing the form, you don't see values in the combo box. Values appear when the form is running and the user is setting or entering values as shown below.

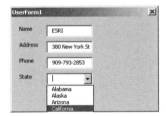

As you add controls to a form, you set their properties in the Visual Basic Editor Properties window. Properties determine the appearance of a control, including its height, width, color, and text. Below, the Add button's BackColor property has been set to green and its ForeColor property to white. The word "Add" has been typed in for the Caption property.

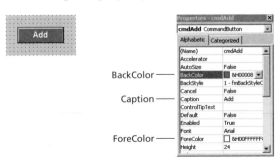

BackColor

Caption

ForeColor

After setting the properties for a control, you then write its VBA code. Controls have procedures that are coded to run when the control is used. For example, when a user clicks the Add button, your coded procedure runs to add a record into a database table.

Since a dialog box will fill a particular need for collecting data or performing a task, it is helpful to begin its design by talking to the people who will use it. For the moment, suppose that these people are city workers who repair streetlights. They have laptops in their trucks and use ArcMap to locate light poles. After a repair, they write the repair record on a paper form, which is given to someone in the office to enter into a database. Workers could save time if they had an ArcMap form to use in the field.

By examining the existing system of getting repair data into the database, you can determine the information that gets collected (repair date, type of repair, light identification number) and the tasks that are performed (add a record into the database for each repair).

Before making the form with Visual Basic Editor, sketch it out on paper to show to the work crews and data entry staff to see if it contains all the controls that they would expect.

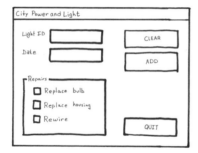

Once your paper design receives everyone's approval, you can start Visual Basic Editor, open a blank form, and drag controls onto it from the Toolbox. Then for each control you can set properties to get it to look like the paper design.

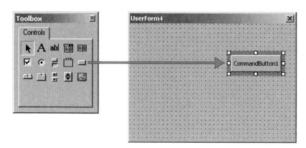

Exercise 3

When Parcel Viewer users click the Calculate Tax button (created in exercise 2c) and enter a parcel value, the tax that is calculated and displayed is the residential rate of 2 percent. However, while most parcels in the city are zoned residential, some are zoned commercial or industrial and each zone has a different tax rate. To calculate taxes more accurately, users need to be able to choose their parcel's zoning code.

In this exercise, you will create a tax calculator dialog box that looks like the sketch below. It will contain a TextBox to enter a parcel value, a ComboBox with a list of zoning values, and a Label for the calculated tax amount. In chapters 4 and 5, you will write the VBA code to make it all work.

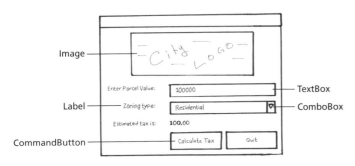

1. Start ArcMap and open **ex03a.mxd** in the **C:\ArcObjects\Chapter03** folder.

Only the Main Menu and Parcel Viewer toolbars need to be turned on for this exercise.

2. Click the Tools menu, point to Macros, and click Visual Basic Editor (or press Alt + F11).

You are going to create the new form in the project for ex03a.mxd.

3 In the Project window, right-click Project (ex03a.mxd), point to Insert, and click UserForm.

A new form called UserForm1 opens, along with the Toolbox of controls.

 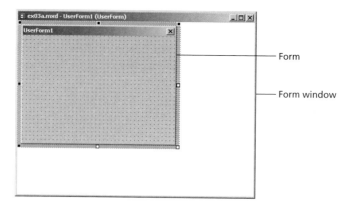

Form

Form window

When you create forms and controls, VBA assigns them a variable name. For example, if you were to write code with the new form above, you would refer to it as UserForm1. You will write that kind of code in the next chapter.

Since the name UserForm1 isn't very meaningful, you will change it by setting the Name property in the Properties window.

4 In the Properties window, type **frmTax** for the Name property. Press Enter.

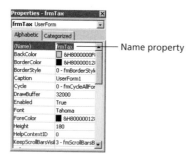

Name property

Next you will use the Properties window to change the form's gray color to white.

5 Click the BackColor property.

The property highlights and a drop-down arrow appears.

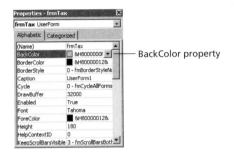 —— BackColor property

6 In the BackColor property, click the drop-down arrow and click the Palette tab. Click the white square in the upper left corner of the color palette.

Click white ——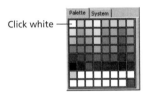

You see the form change color from gray to white.

7 In the Properties window, set the following form properties by clicking on each one and typing the listed value.

— For Caption, type **Tax Calculator**

— For Height, type **192.75**

— For Width, type **255.75** and press Enter

The numbers you entered for height and width are in pixels. The size of the form and its controls are all given in pixel units.

Next you will add a city logo to the form and set its properties.

8 Click the form window to make it active.

9 If the Toolbox has closed, reopen it.

Sometimes, all you need to do to reopen the Toolbox is make the form window active. If the form window is active and the Toolbox is still not open, click the Toolbox button on the Standard toolbar.

10 Drag an Image control from the Toolbox to the form.

Don't worry about where you put the control on the form because you will set its position in the following step.

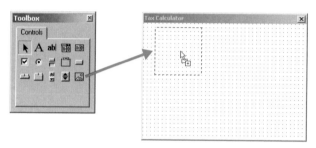

11 In the Properties window, set the following properties for the Image control.

The Top and Left properties that you will set below refer to the coordinates on the form where the upper left corner of the image will be placed. Like Height and Width, the Top and Left properties are set in pixel units.

— For Name, type **imgLogo**

— Click BackColor, click the drop-down arrow, click Palette, and click white

— Click BorderColor, click the drop-down arrow, click Palette, and click white

— For Height, type **60**

— For Left, type **18**

— For Top, type **6**

— For Width, type **216** and press Enter

The image control is resized and its color is white.

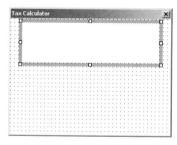

Next you will specify the folder location of an image file to draw in the rectangle.

12 In the Properties window, click the Picture property and click the Ellipsis button that appears.

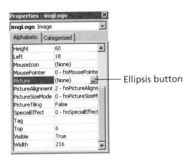

— Ellipsis button

13 In the Load Picture file browser, navigate to **C:\ArcObjects\Data\Manhattan_KS** and click **Logo.jpg**. Click Open.

The picture needs to be shrunk so you can see the entire logo.

To shrink it, you will set the Image control's PictureSizeMode property.

14 In the Properties window, click the PictureSizeMode property and click the drop-down arrow.

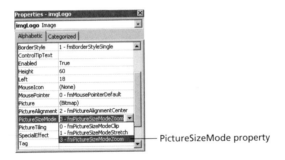

— PictureSizeMode property

15 Click 3-fmPictureSizeModeZoom.

The fmPictureSizeModeZoom option forces the picture to fit inside the rectangle without stretching or distortion. The other options either clip or stretch it.

There is an apple in the logo because Manhattan is nicknamed the Little Apple.

Next you will add a text box, a combo box, and a label. Users will enter parcel values into the text box, the combo box will display a list of zoning choices, and the label will be used to display the calculated tax amount.

16 Make the form window active and open the Toolbox. From the Toolbox, drag a TextBox, ComboBox, and Label and drop them anywhere on the form.

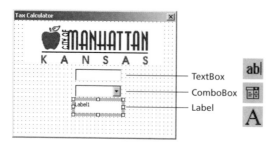

17 Select each control, one at a time, and set the following properties for each.

TextBox
— Name: **txtParcelValue**
— Height: **18**
— Left: **96**
— Top: **78**
— Width: **150** and press Enter

ComboBox
— Name: **cboZoning**
— Height: **18**
— Left: **96**
— Top: **102**
— Width: **150** and press Enter

Label
— Name: **lblTaxAmount**
— Height: **18**; Width: **72**
— Left: **96**
— Top: **126** and press Enter
— Caption: (remove all text to make the caption blank)
— Click Font and click the Ellipsis button. In the Font dialog box, click Bold for Font Style. Click OK.

You see the three controls with the properties applied. They're all aligned because you set their Left property to 96.

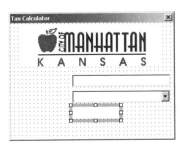

18 From the Toolbox, add three labels. Position them as shown below.

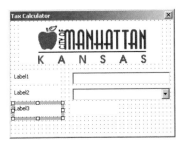

19 In the Properties window, set the Name and Caption properties for the three labels as follows.

— Name: **lblValue**; Caption: **Enter parcel value:**

— Name: **lblZoning**; Caption: **Zoning type:**

— Name: **lblTax**; Caption: **Estimated tax:**

The purpose of these three labels is to display descriptive text next to the Parcel Value text box, Zoning combo box, and Tax Amount label. As they are just labels for the other controls (they don't *do* anything), you won't be writing any code for them.

Next you will select all three labels to right-justify them.

20 Drag a box around all three labels.

Three labels selected →

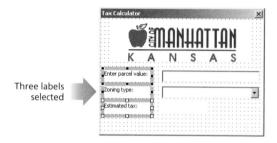

21 In the Properties window, click the TextAlign property, click the drop-down arrow, and click 3-fmTextAlignRight.

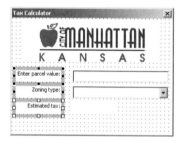

You will now complete the form by adding a button to display the tax and a button to close the dialog box when the user is done.

22 From the Toolbox, drag two CommandButtons to the form and position them as shown below.

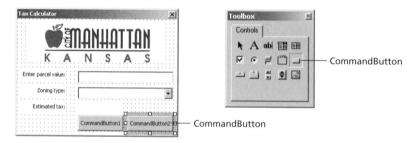

CommandButton

CommandButton

23 For each CommandButton, set the following properties.

Tax button (on the left)
— Name: **cmdCalculateTax**
— Caption: **Calculate Tax**

Quit button (on the right)
— Name: **cmdQuit**
— Caption: **Quit**

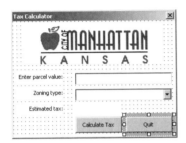

The form has all its controls and is ready for a test run.

24 Click the form's window to make it active. Click the Run Sub/User Form button.

The form appears as a dialog box.

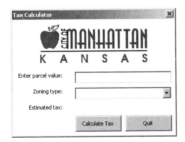

No code runs, because you haven't written any. You do a test run to see how the dialog box will look to the user, without any selected controls and without any grid dots.

25 Click the × in the upper right corner of the dialog box to close it.

Normally, you would also be able to click the Quit button on the dialog box, but since you haven't added any VBA code, it doesn't work yet.

26 Close Visual Basic Editor.

27 If you want to save your work, click the File menu in ArcMap and click Save As. Navigate to **C:\ArcObjects\Chapter03**. Rename the file **my_ex03a.mxd** and click Save. If you are continuing with the next chapter, leave ArcMap open. Otherwise close it.

Programming with objects

Programming with methods

Getting and setting an object's properties

You worked with objects in the previous chapter when you made a user form and added controls to it. The form is an object and so are its controls. (These are VBA objects, not ArcObjects. They come with any application that includes VBA.) You didn't do any programming, however, so the form and control objects don't work yet. To get these objects to do what you want, you write code for their events, methods, and properties.

An *event* is a user action (like a mouse click) that happens to an object. An event procedure is code that runs when the action occurs. You worked with events and event procedures in chapter 2 when you created the Help and Tax Calculator UIButtons and coded their click event procedures. When someone clicks either button, your code runs.

A *property* is a characteristic or an attribute of an object. In a way, properties are like variables because they both store a value that you can change. You set an object's properties to make it look different from other objects. For instance, if you had some otherwise identical buttons, you could tell them apart by setting each one's Caption property differently.

In the previous exercise, you used the Properties window to set property values. For example, after you added the cboZoning combo box to the form, you set its name, left, top, and width properties.

An object's property values can also be set by writing lines of code that look like a sentence mixed with a math equation. This programmer's grammar, shown in the sample code that follows, is called the "object dot property" syntax. To set an object's property with a line of code, you begin with the object's name, a dot, and the property you want to set. Then you use the equals sign and the value you want to set the property to. It's similar to setting a variable's value.

```
Object.Property = "SomeValue"
```

Instead of using the Properties window to set the cboZoning combo box's left, top, and width properties, you could set them with the following three lines of code:

```
cboZoning.Left = 96
cboZoning.Top = 102
cboZoning.Width = 150
```

Methods, also called behaviors, are the things that an object can do. A form object, for example, has the following methods: Copy, Cut, Hide, Move, Paste, PrintForm, and Show (to name a few). Each method is a block of code that runs when called into action.

The syntax for calling a method into action is:

```
Object.Method
```

Suppose you have a form object called frmAddRecord and you want to run its Show method. (The Show method opens the given form.) You would write the following line of code:

```
frmAddRecord.Show
```

When the line of code runs, the form opens to the user.

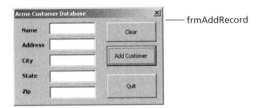 ——— frmAddRecord

Methods are like events in that calling them into action causes a block of code to run. The difference is that while event procedures are empty until you write the code for them, the procedures that go with methods have already been written for you.

For the time being, you will only call methods into action. In chapter 9, you will learn how to write the code for a method.

In the next two exercises, you'll work first with methods, then with properties.

CHAPTER 4 • PROGRAMMING WITH OBJECTS

Programming with methods

The captain of the spaceship Atlantis sits on the bridge and gives orders: beam up some crew members, put the deflector shields down, go to warp speed. If you think of the spaceship as an object, then the various orders it carries out are its methods. For example,

```
Atlantis.WarpSpeed
```

is the method that makes the spaceship go faster than light. Of course, going to warp speed isn't a one-step task. A whole series of things takes place—people in the engine room flip switches, push levers, and monitor temperature levels; other people in navigation check the route and locate obstacles.

A method, in other words, entails a procedure—a list of things to do. It may be a long list or it may be a short list, but either way you always use the object.method syntax to call it into action.

Some methods have variations that you specify with arguments you add after the method. The spaceship's Shields method, for example, has an argument to control the shield status, which can be either up or down. The status argument has two settings. To put the shields down you would write:

```
Atlantis.Shields Down
```

Other methods, like the BeamUp method, have multiple arguments. When you write a line of code with arguments, you separate them with commas.

```
Atlantis.BeamUp Andrew, Thad, Michael
```

A page or two ago, you learned that methods are procedures that have already been coded for you. You may be wondering—by whom? The methods that belong to VBA objects, like forms and controls, were coded by Microsoft programmers. The methods for GIS objects, like UIControls, maps, and layers, were coded by ESRI® programmers. Regardless of who does the coding—even if it's you—methods are always called with the same object.method syntax in VBA.

Exercise 4a

In chapter 3, you created the Tax Calculator dialog box by dragging controls onto it and setting their properties. However, you didn't write any code there. In this exercise you are going to write code to get the dialog box to work.

Your first task will be to write code for the Calculate Tax button on the Parcel Viewer toolbar, so users can click this button to open the dialog box. Next, you will write code for the dialog box's Quit button, so users can click Quit to close the dialog box.

Finally, you will write code to add the words Residential, Commercial, and Industrial to the zoning combo box. These words will become selectable choices on the combo box's drop-down list.

 —— cboZoning

1 Start ArcMap and open **ex04a.mxd** in the **C:\ArcObjects\Chapter04** folder.

When the map opens, you see the Manhattan city parcels and the Parcel Viewer toolbar.

2 On the Parcel Viewer toolbar, right-click the Calculate Tax button and click View Source.

The ThisDocument code module opens. You see several procedures that you coded in previous exercises. However, the Calculate Tax button's click event procedure is empty. In exercise 2c, you wrote code that used an input box to do tax calculations. Since you created the Tax Calculator dialog box, that code has become obsolete, so it was deleted for you.

In chapter 3, you created the Tax Calculator dialog box and named it frmTax. This is the name that you will use in your code to refer to the Tax Calculator form. In the next line of code, you will use the Show method to open the form.

3 In the Calculate Tax click event, indent for a new line of code and type **frmTax** and a dot.

```
frmTax.
```

After you type the dot, a drop-down list of the form object's properties and methods appears (unless you've turned the Auto List Members option off under Tools > Options).

Icons to the left of each item indicate whether it is a property or method. Properties look like a hand and finger pointing at a database table. Methods look like a flying green brick.

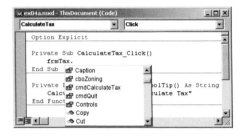

4 Scroll down in the list and double-click the Show method. (You could also type **Show**.)

When this line of code runs, the Show method opens the tax form.

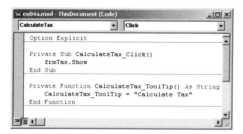

Now that you have programmed a way for users to open the Tax Calculator, you will also program a way for them to close it. You will add code to the dialog box's Quit button.

5 In the Project window, under Project (ex04a.mxd), double-click frmTax in the Forms folder to open the Tax Calculator form.

6 On the form, right-click the Quit button and click View Code.

The form's code window opens and you see the wrapper lines for the Quit button's click event procedure.

7 In the click event, add the following line of code.

```
frmTax.Hide
```

Again, you are using the object.method syntax, where frmTax is the object and Hide is the method. When a user clicks the Quit button, this code will run.

Next, you will write code to add the zoning names to the zoning combo box. The code will be written in the form's initialize event procedure. In the following steps, you will navigate to that procedure and write code in it.

8 In the code window for frmTax, click the object list drop-down arrow and click UserForm.

The object list contains the names of each object (control) on the form, as well as the form object itself. You might expect to see frmTax in the list—since that's what you named the form—but instead you see UserForm. No matter what name you give the form, VBA always displays it as UserForm in the object list.

Before you have a chance to do anything, Click is selected in the Procedure drop-down list and its wrapper lines are added in the code window.

Object list

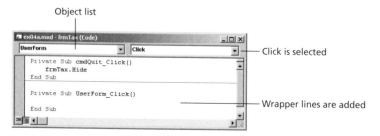

Click is selected

Wrapper lines are added

Click is the form's default event procedure. When you select an object in the object list, its default event procedure's wrapper lines are automatically added (unless they are already there). You are not going to code the UserForm's click event, so ignore these lines or remove them.

With UserForm selected in the object list, if you click the drop-down arrow in the procedure list you'll see the rest of the UserForm's event procedures. You will add its initialize event procedure next.

9 Click the procedure list drop-down arrow and click Initialize.

The initialize event wrapper lines are added.

Procedure list

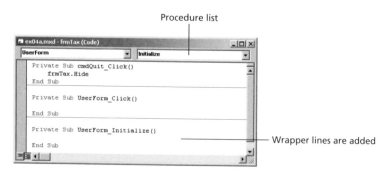

Wrapper lines are added

Normally, you can set an object's properties with Visual Basic Editor's Properties window. However, combo boxes don't have a property to hold the values that appear in their drop-down lists. For this, you have to write code.

You add values to a combo box's list with the AddItem method. For example, to add the color choices Red, Green, and Blue to a combo box called cboColor, you would write the following three lines of code:

You put these three lines of AddItem code in a form's initialize event, because code in the initialize event runs in the moments before the form opens to the user. That way, the combo box is filled with choices before the user sees the form. When the user clicks the drop-down list, the choices are there and ready to be selected.

10 In the UserForm's initialize event, add the following three lines of code.

The AddItem method adds one value to the combo box's drop-down list. To add three values, you have to use the AddItem method three times.

```
cboZoning.AddItem "Residential"
cboZoning.AddItem "Commercial"
cboZoning.AddItem "Industrial"
```

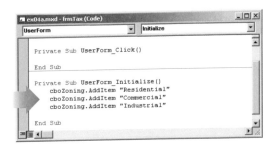

11 Close Visual Basic Editor.

You have written procedures to help the user open and close the Tax Calculator dialog box and to add values to the zoning combo box. Next you will test all three procedures.

12 On the Parcel Viewer toolbar, click Calculate Tax.

The form's Show method runs and the dialog box opens.

13 Click the Zoning type drop-down arrow and click Industrial.

The zoning types appear in the list and can be selected. You see the choices in the list, because they were added as the form initialized. (These choices don't actually do anything yet, because there is no code behind them.)

14 Click Quit.

The form's Hide method runs and the form closes.

In the next exercise, you will continue to code the dialog box.

15 If you want to save your work, click the File menu in ArcMap and click Save As. Navigate to **C:\ArcObjects\Chapter04**. Rename the file **my_ex04a.mxd** and click Save. If you are continuing with the next exercise, leave ArcMap open. Otherwise close it.

Getting and setting an object's properties

In the Properties window below, you see properties for the Quit command button. The caption property is, naturally enough, the word Quit. Naturally, that is, if you speak English. If your users are Spanish speakers, you would want the button to say Terminar instead. And you could easily make this change just by typing Terminar where Quit is now.

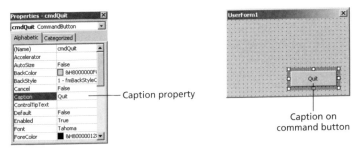

— Caption property

Caption on
command button

But suppose your users included both English and Spanish speakers. You might want to create a bilingual dialog box in which captions switch from one language to another depending on who's using the application. To do that, you would first have to write some code that asks the user for their language preference and stores the value in a variable. You would then go on to set the appropriate caption property with a line of code.

The code to set a caption property would look something like this:

```
cmdQuit.Caption = "Terminar"
```

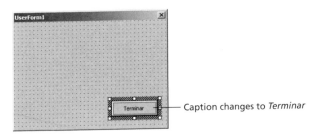

— Caption changes to *Terminar*

In this example, you were concerned with *setting* the value of a property. In other situations, you may want to find out the value of a property that has already been set. It may be a property that changes according to user input, and you need to know what it is so that you can use it in another calculation. Finding out a property's value and storing it in a variable is called *getting* a property.

Suppose you're writing a handy little tool to convert feet to meters. Your dialog has a few different controls—a couple of text boxes and a button. The top text box is called txtFeet and its Text property is set to whatever value the user types in. The bottom text box is called txtMeters and its Text property will be set programmatically after the user clicks the Convert to Meters button.

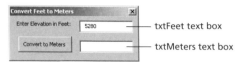

To get a property, you use a variable and an assignment statement with the following syntax:

```
variable = object.property
```

Here, you want to get the Text property of the txtFeet object. In the next line, the variable is called Feet, but it could be called anything.

```
Feet = txtFeet.Text
```

After the line runs, the variable will contain the value typed in by the user (5280 in this example).

Now, where does this line of code go exactly? It goes inside the click event procedure of the Convert to Meters button.

```
Public Sub cmdConvertToMeters_Click()
    Feet = txtFeet.Text
    txtMeters.Text = Feet * 0.3048
End Sub
```

When the user clicks the Convert to Meters button, the first line of code *gets* the Text property from the txtFeet object, assigning this value to the variable Feet. In the second line of code, to the right of the equals sign, the variable value is multiplied by the conversion factor. The line as a whole *sets* the result as the Text property of the txtMeters object.

When the event procedure runs, the value (1609) displays in the txtMeters text box.

An experienced programmer could bypass the explicit variable assignments and write the code more efficiently in a single line: txtMeters.Text = txtFeet.Text * 0.3048.

Exercise 4b

In this exercise, you will write code for the Tax Calculator dialog box to get the user's parcel value, calculate its tax, and display that amount.

1 Start ArcMap and open **ex04b.mxd** in the **C:\ArcObjects\Chapter04** folder.

When the map opens, you see the Manhattan city parcels and the Parcel Viewer toolbar.

2 Click the Tools menu, point to Macros, and click Visual Basic Editor.

3 In the Project window, under Project (ex04b.mxd), open the Forms folder, and double-click frmTax to open the Tax Calculator.

4 On the form, right-click the Calculate Tax button and click View Code.

The form's code module opens to the button's click event wrapper lines. You also see the events you added in the previous exercise.

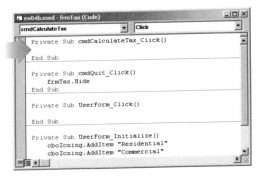

You are about to create a variable to store the parcel values and you need to choose its data type. The Currency data type can store values up to 922 trillion, but requires 8 bytes. The Long data type can store values up to 2.1 billion and takes up just 4 bytes. Since all the parcels in your area are well under 2.1 billion, you'll use Long.

5 In the Calculate Tax click event procedure, add two lines of code to declare and set a Long variable to hold the user's parcel value. To get the parcel value, use the text box's Text property.

```
Dim userValue As Long
userValue = txtParcelValue.Text
```

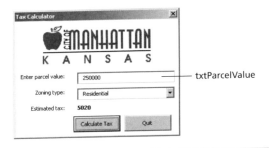

txtParcelValue

6 Add two more lines of code to declare and set a variable to calculate and hold the tax amount.

```
Dim taxAmount As Long
taxAmount = (userValue * 0.02) + 8.55 + 11
```

This is the same equation you used in chapter 2 when you created the Tax Calculator button. There is a 2-percent residential zoning charge (0.02), an $8.55 convention center fee, and an $11.00 fire truck fee.

Instead of 0.02, it would be ideal to use a variable that changes to represent residential, commercial, or industrial tax rates. In the next chapter, you'll learn how to do this as you finish the Tax Calculator dialog.

7 Just above the two lines of code you added in the previous step, insert two comment lines to remind yourself and other programmers that you have assumed the residential zoning rate until actual zoning values can be determined.

Comments begin with an apostrophe and appear green in the code window. Since VBA ignores comments, they won't slow your code.

```
'Zoning values will be determined later.
'Residential tax rate is assumed in calculation below.
```

Next you'll add a line of code to display the tax amount. You'll do it by setting the Caption property of the lblTaxAmount label, shown below with a value of 5020.

8 At the end of the event procedure, add the following line of code to set the label's caption property equal to the calculated tax amount.

```
lblTaxAmount.Caption = taxAmount
```

The Quit button's click event uses the Hide method to close the dialog box. Hiding a dialog box keeps it, its controls, and all their property settings in memory. When a hidden dialog box is reopened, the UserForm initialize event procedure doesn't run again. That means its most recent settings reappear. Were a user to select Industrial in the zoning combo box and then click Quit, Industrial would appear when the dialog box was reopened.

You will write code in the Quit button's click event to clear the combo box's selection area, the text box's text area, and the label's caption, so that they will be empty each time the dialog box is reopened.

9 In the cmdQuit_Click event, add the following three lines of code.

This code sets the Text property of the combo box and text box and the Caption property of the label to a blank text string.

```
cboZoning.Text = " "
txtParcelValue.Text = " "
lblTaxAmount.Caption = " "
```

The code is ready to test.

10 Close Visual Basic Editor.

11 In ArcMap, on the Parcel Viewer toolbar, click Calculate Tax.

12 For parcel value, type **250000**.

13 For Zoning type, click the drop-down arrow and click Residential.

Your coded calculation uses the Residential tax value of 2 percent, so even if you select Commercial or Industrial the same value will be calculated.

14 Click Calculate Tax.

The estimated tax, 5020, appears.

Next you will test to see that the parcel value, zoning type, and estimated tax are cleared when the dialog box is closed and reopened.

15 Click Quit.

16 On the Parcel Viewer toolbar, click Calculate Tax.

As the form opens, its initialize event does not run. The code that you added to the Quit button's click event clears the parcel value text box, the zoning combo box's selection area, and the tax amount label.

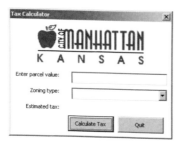

17 Click Quit.

18 If you want to save your work, click the File menu in ArcMap and click Save As. Navigate to **C:\ArcObjects\Chapter04**. Rename the file **my_ex04b.mxd** and click Save. If you are continuing with the next chapter, leave ArcMap open. Otherwise close it.

Code for making decisions

Making a Case for branching

Coding an If Then statement

Unless otherwise directed, lines of code in a procedure run, in order, from first to last. However, lines of code can be grouped into blocks, and decision-making statements, called Case and If Then, can control which blocks run and which don't.

Case statements handle multiple-choice situations. Suppose you have a dialog box that asks "What layer do you want to add to the map?" A combo box after the question contains five choices: Roads, Rivers, Lakes, Soils, and Elevation. If the user picks Roads, a block of code runs to add the roads layer to the map. If the user picks Lakes, a different block of code runs to add the lakes layer. Each choice causes different code to run.

In a Case statement, you write a block of code to run for each possible choice. Depending on the choice, the appropriate block (and only that block) runs.

If Then statements handle true-false situations. Suppose you have a dialog box that asks "Do you want to print the map?" and contains a Yes button and a No button. If the user clicks Yes, one block of code runs and a map is printed. If the user clicks No, no code runs and no map prints.

In an If Then statement, you write two blocks of code. One runs when the statement is true and the other runs when the statement is false.

Making a Case for branching

Case statements process multiple-choice situations just like you do when you approach a traffic signal: when the light is red, you stop; when the light is green, you proceed with caution; when the light is yellow, you slow down and stop unless you can get into the intersection before the light turns red. When the light is flashing red, or when the lights are out, you treat the intersection as if it had a stop sign.

A Case statement starts with the Select Case keywords and a variable that contains a value, which for the status of a traffic light might be red, green, or yellow. For each possible value, there is a Case keyword and the value. After that, you add all the code necessary for that value. The End Select keywords end the statement.

```
Select Case theTrafficSignalValue

    Case Red

        Stop the car

    Case Green

        Continue with caution

    Case Yellow

        Stop the car if you can't make it before red

End Select
```

The Case statement above assumes that theTrafficSignalValue has only three possibilities: red, green, or yellow. But if there's a power failure in the area, the lights won't work. To account for any odd or unexpected values, you can use the Else keyword. Below, Case Else is a fourth case added to handle a traffic signal value other than red, green, or yellow.

```
Case Else
    Treat the intersection as if it has a stop sign
```

Exercise 5a

Parcels in the city have one of three zoning types, residential, commercial, or industrial, and each zoning type has a different tax rate. In this exercise, you'll create a Case statement to determine the user's zoning type and apply the appropriate tax rate.

1 Start ArcMap and open **ex05a.mxd** in the **C:\ArcObjects\Chapter05** folder.

When the map opens, you see the Manhattan city parcels and Parcel Viewer toolbar.

2 Click the Tools menu, point to Macros, and click Visual Basic Editor.

3 In the Project window, under Project (ex05a.mxd), double-click frmTax under Forms to open it.

You see the Tax Calculator form.

4 On the form, double-click the Calculate Tax button to open the form's code window.

Double-clicking a control opens the form's code window (or brings it to the front if it's already open). It opens to the control's default event procedure, which, for the Calculate Tax CommandButton, is the click event.

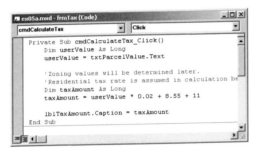

Now you'll add some code at the top of the click event procedure to include the zoning type in the calculation. You'll need a variable to hold the tax rate (which changes) and a Case statement that assigns a value to this variable according to the user's selection in the zoning combo box. If the user picks Residential, the variable value will be 0.02; if they pick Commercial, it will be 0.023; and if they pick Industrial, it will be 0.0275.

5 Insert a new line between the Private Sub line and the first Dim statement.

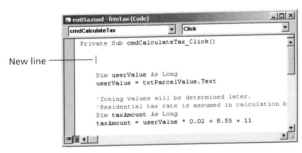

New line

6 On the new line, add the following code to declare a variable for the tax rate.

```
Dim sngTaxRate As Single
```

The Single data type only requires four bytes of space and can hold decimal values. It's the most efficient numeric data type for storing small numbers that have decimal places.

Next, you'll code the Case statement. In exercise 4a, you used the AddItem method to populate the combo box with choices. When the user makes one of these choices, it is set as the combo box's Value property. The possible values of that property determine the cases you need to code.

Making a Case for branching

In each branch of the Case statement, sngTaxRate will be set to the tax rate for the selected zoning type. (The Industrial type is selected in the graphic below.)

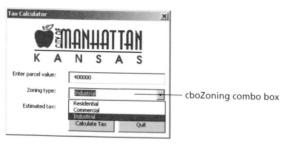

cboZoning combo box

7 After the line of code declaring the sngTaxRate variable, add the first and last lines of a Case statement. For the Case statement's value, use cboZoning.Value.

```
Select Case cboZoning.Value

End Select
```

Next you will add three blocks of code, one for each type of zoning.

8 Inside the Case statement, add a Case to check for Residential value. Add a second line of code to set the tax rate variable to 0.02.

```
Case "Residential"
    sngTaxRate = 0.02
```

Each case is indented from the main Select Case statement and each case's block of code is indented to make it easy to read.

9 Add two more Cases to check for Commercial and Industrial values and set the tax rate variable for each.

```
Case "Commercial"
    sngTaxRate = 0.023

Case "Industrial"
    sngTaxRate = 0.0275
```

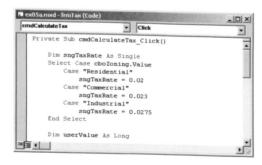

10 Scroll down in the code to find the two comments about residential zoning and delete them.

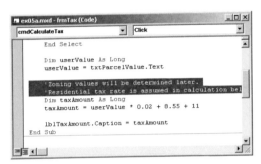

11 Locate the line that calculates the tax amount. Replace 0.02 with the variable, sngTaxRate.

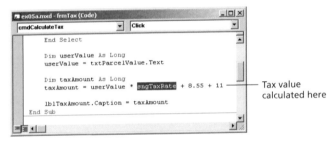

Tax value calculated here

The code is now ready to test.

12 Close Visual Basic Editor.

13 On the Parcel Viewer toolbar, click the Calculate Tax button.

14 For parcel value, type **400000**.

15 For Zoning type, click the drop-down arrow and click Industrial.

16 Click the Calculate Tax button.

The estimated tax, 11020, appears. To test your Case statement with the other zoning types, you will switch zoning types and recalculate the tax.

17 For Zoning type, click the drop-down arrow, and click Commercial. Click Calculate Tax.

The estimated tax, 9220, appears.

18 For Zoning type, click the drop-down arrow, and click Residential. Click Calculate Tax.

The estimated tax, 8020, appears.

Depending on the user's zoning, your Case statement runs the appropriate block of code to calculate the tax rate.

19 Click Quit.

20 If you want to save your work, click the File menu in ArcMap, and click Save As. Navigate to **C:\ArcObjects\Chapter05**. Rename the file **my_ex05a.mxd** and click Save. If you are continuing with the next exercise, leave ArcMap open. Otherwise close it.

Coding an If Then statement

Out driving, you approach a fork in the road where going left takes you into the city and right takes you into the countryside. You need to make a decision on which way to turn and to do that you ask yourself a true-or-false question. Is today a work day? If it's a work day you go left to your office in the city. If it's not a work day, you go right and take a relaxing drive in the countryside.

You can write code to process this decision with the If Then statement below. The first line in an If Then statement contains an expression (below, it's Today = aWorkDay) between the keywords If and Then. If the expression is true, the block of code between If Then and the Else keyword runs. If the expression is false, the block of code between Else and End If runs. The End If keywords end the statement.

```
If Today = aWorkDay Then

    Turn left to work in the city

Else

    Turn right to drive in the countryside

End If
```

The essential thing about If Then expressions (also called Boolean or logical expressions) is that they are either true or false. For example, 4 < 5, 20 > 40, 2 <= 2, and "Hello" = "Good bye" are all logical expressions, because when you evaluate their logic they result in either a true or false answer. But 4 + 5, on the other hand, is not a logical expression. When you evaluate 4 + 5, you get 9, and the number 9 is not the same thing as true or false.

Logical expressions can compare two values using math symbols called comparison operators. These include the equals sign (=), greater than (>), less than (<), greater than or equal to (>=), less than or equal to (<=), and not equal to (<>). In the expression below, if x is 10 and y is 5, the expression is true. If x is 2 and y is 15, the expression is false.

```
x > y
```

Logical expressions can be combined with logical connectors, which are English words like AND and OR. Below, AND combines two expressions into one larger expression. In order for the full expression to be true, both smaller expressions must be true.

```
x > y AND a = b
```

With OR, only one of the expressions must be true for the full expression to be true.

```
x > y OR a = b
```

Another way to create an expression is with a function. VBA comes with a variety of predefined functions including a group that are named with the Is prefix: IsDate, IsEmpty, IsError, IsMissing, IsNull, IsNumeric, and IsObject. The Is functions all result in true or false. For example, the IsNumeric function tests a variable to see if it contains a number. If x contains a number, the expression below evaluates to true; otherwise, it's false.

```
If IsNumeric(x) Then
```

A third way to create an expression is by getting an object's property. CommandButtons have many true or false properties, as shown in the Properties window below.

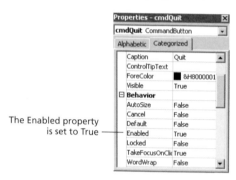

The Enabled property is set to True

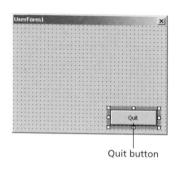

Quit button

You can use the object.property syntax to create an expression. When the Quit button's Enabled property is True, the following expression is true:

```
If cmdQuit.Enabled Then
```

An If Then statement can evaluate multiple expressions with the ElseIf keyword. Below, two ElseIfs are used to evaluate expressions for alkaline and neutral pH levels. If the If Then expression is true, or if either of the ElseIf expressions are true, the code after that expression runs. If a non-pH value is entered (less than zero or greater than 14), the code after the Else keyword runs.

```
If intPHLevel < 7 And intPHLevel >= 0 Then
    MsgBox "You have an acid"
ElseIf intpHLevel > 7 And intPHLevel <= 14 Then
    MsgBox "You have an alkaline"
ElseIf intpHLevel = 7 Then
    MsgBox "The value is neutral"
Else
    MsgBox "The value is outside the pH scale"
End If
```

Exercise 5b

If users enter nonnumeric values in the Parcel Value text box, they get a type mismatch error when they click the Calculate Tax button. Below, the user types "$200,000" (with quotation marks) into the text box, clicks Calculate Tax, and gets an error message.

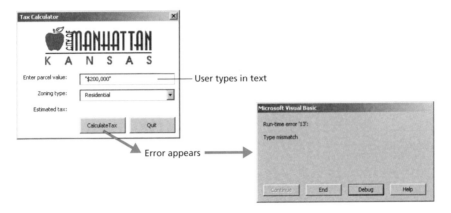

The error appears because the quotation marks make $200,000 a string instead of a number. VBA cannot multiply a string and a number:

```
"$200,000" * 0.02
```

In this exercise, you will code the Parcel Value text box's Change event. Each letter or number that the user types into the text box causes code in the Change event procedure to run. Whenever the user makes a change to the text box, the code runs.

You'll put code in there to determine if users are typing letters or numbers. If they type letters, your code will disable the Calculate Tax button by setting its Enabled property to false. When a button is disabled, it is grayed out.

Enabled = False —— Calculate Tax

1. Start ArcMap and open **ex05b.mxd** in the **C:\ArcObjects\Chapter05** folder.

When the map opens, you see the Manhattan city parcels and the Parcel Viewer toolbar.

2. Click the Tools menu, point to Macros and click Visual Basic Editor.

3 In the Project window, under Project (ex05b.mxd), double-click frmTax under Forms to open it.

You see the Tax Calculator. You will write code that validates the values that are typed into the Parcel Value text box.

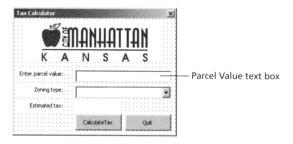

Parcel Value text box

4 On the form, double-click the Parcel Value text box (txtParcelValue).

The form's code module opens to the text box's change event procedure. Change is the default event for text boxes and it's the one you will write code in.

Change event

5 Inside the change event, start an If Then statement. For the logical expression, use the IsNumeric function to evaluate the Parcel Value text box's Text property.

```
If IsNumeric(txtParcelValue.Text) Then
```

The IsNumeric function is true when the user enters a number and false when any other values are entered.

6 Indent and add the following line to set the Calculate Tax button's Enabled property to true.

```
cmdCalculateTax.Enabled = True
```

As long as the user types numbers into the Parcel Value text box, the Calculate Tax button will be enabled.

7 Outdent and add the Else statement.

```
Else
```

8 Indent and add the following line to set the Calculate Tax button's Enabled property to false.

```
cmdCalculateTax.Enabled = False
```

When the user types anything other than a number into the Parcel Value text box, the Calculate Tax button will be disabled (grayed out).

9 Finish the If Then statement with the End If keywords.

```
End If
```

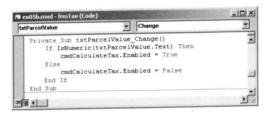

The If Then statement is ready to test.

10 Close Visual Basic Editor.

11 On the Parcel Viewer toolbar, click the Calculate Tax button.

12 For the parcel value, type **Hello**.

As soon as you type any text, Calculate Tax is disabled.

13 Change the parcel value to **250000**.

As soon as you type numbers, the Calculate Tax button becomes enabled.

14 Select Residential zoning and click Calculate Tax to see the button work correctly.

The estimated tax of 5020 displays.

15 Click Quit.

You type a lot of variable, method, and property names. To help minimize typing mistakes, you can let VBA's code completion option type for you. For example, you type

```
cmdC
```

as if you are going to type **cmdCalculateTax**.

But after typing just **cmdC**, you press Contol + spacebar on the keyboard. This calls VBA's code completion option into action. It finishes the typing for you:

```
cmdCalculateTax
```

Use code completion to reduce typing mistakes in variable, method, and property names.

16 If you want to save your work, click the File menu in ArcMap, and click Save As. Navigate to **C:\ArcObjects\Chapter05**. Rename the file **my_ex05b.mxd** and click Save. If you are continuing with the next chapter, leave ArcMap open. Otherwise close it.

Using subroutines and functions

Calling a subroutine

Passing values to a subroutine

Making several calls to a single subroutine

Returning values with functions

Up to this point, you've been working mostly with event procedures, which run in response to a user action or a change in the system state. In this chapter, you'll take a closer look at subroutines and functions.

Like an event procedure, a subroutine is a list of instructions that carries out a task. The task may be to print a map, buffer a feature, add a field to a table, or any number of things. What distinguishes a subroutine from an event is the cue that sets it in action. An event procedure runs when its event occurs. A subroutine runs when it is *called* by another procedure. A call is a line of code that says, "Procedure So-and-so, it's your turn to run."

The calling procedure may be an event or it may be another subroutine. For example, an event procedure could call Subroutine A, which calls Subroutine B, which calls Subroutine C, and so on.

Like a subroutine, a function is a procedure that waits to be called. It's different from a subroutine in this respect: when it's finished, it returns a value to the line of code that called it. A subroutine might assign symbology to world countries based on their population values. A function might sum the population value for each country and return the earth's population.

Instead of having lots of different procedures calling each other, you might be wondering why you couldn't put them all into one big block. In the example above, why not take all the code for Subroutines A, B, and C and put it right inside the event procedure? That way, when the event occurs, all the code runs and nobody has to call anybody.

In fact, you could write code that way—it's just not efficient. Writing code in discrete blocks that call other discrete blocks has several advantages. For one thing, it makes it easier to reuse the code. Suppose you've made three buttons with different click event procedures. One draws buffer zones, another intersects two layers, and the third uses one layer to clip another. In each case, you want to print the map after the operation. You could copy and paste your map-printing code into all three event procedures, but it's simpler to write the code once and run it with a call from each of the event procedures.

Suppose you decided to do the copy and pasting anyway, and then found that your map-printing code had an error. Instead of debugging it once in a subroutine or function, you'd have to debug it in three different event procedures. The same goes for updating it. Say you want to change the way maps are printed—you'd have to make the change in three places instead of one.

Subroutines and functions also keep your code organized. If you have several tasks to perform in sequence and you code them all in a single long block, it's easy to lose track of what a particular line of code is doing and what has already been done.

In this chapter's four exercises, you will call and modify subroutines and write a function.

Calling a subroutine

You tell a subroutine to run with the Call statement.

```
Public Sub GetMessages()
    Call Message
End Sub
```

The subroutine above is called GetMessages, and it does only one thing: it tells another subroutine, called Message, to run. When called, the Message subroutine runs and displays a message box.

```
Public Sub Message()
    MsgBox "Geography is terrific"
End Sub
```

Any procedure that wants to display the words "Geography is terrific" can call the Message procedure. Below, the DailyQuote subroutine calls Message.

```
Public Sub DailyQuote()
    Call Message
End Sub
```

One procedure can call many others. Below, MakeMap_Click is a button's click event procedure. It tells three subroutines to run. As one finishes, the next one begins.

```
Public Sub MakeMap_Click()
    Call AddCartographicComponents
    Call CheckForPrinter
    Call PrintMap
End Sub
```

A procedure may call a procedure that calls another procedure. Below, cmdMessage_Click calls Test2, which calls Test3.

```
Public Sub cmdMessage_Click()
    Call Test2
    frmGIS.Hide
End Sub

Public Sub Test2()
    Call Test3
    MsgBox "VBA is fun"
End Sub

Public Sub Test3()
    MsgBox "I'd rather be writing procedures"
End Sub
```

The chain of called procedures begins with an event. As each procedure is called, its code runs; when it finishes, control returns to the calling procedure.

Below, comments identify the order in which the lines of code run. To begin with, the first two lines of the click event run. The second line of the click event calls Test2, and its first two lines run. The second line of Test2 calls Test3, and its three lines run. When Test3 finishes, control returns to Test2 and its last two lines run. When Test2 finishes, control returns to the click event and its last two lines run.

```
Public Sub cmdMessage_Click()                            '1
    Call Test2                                           '2
    frmGIS.Hide                                          '10
End Sub                                                  '11

Public Sub Test2()                                      '3
    Call Test3                                          '4
    MsgBox "VBA is fun"                                '8
End Sub                                                 '9

Public Sub Test3()                                     '5
    MsgBox "I'd rather be writing procedures"  '6
End Sub                                                 '7
```

This means that the message "I'd rather be writing procedures" displays before the message "VBA is fun." The "VBA is fun" message displays before the Hide method runs on frmGIS.

Exercise 6a

You work as a programmer on the Washington, D.C., Police Department's Crime Analysis team. Each week, your team meets with the mayor, police chief, and precinct captains to discuss how to reduce crime.

During the meetings, analysts display maps showing where crimes have occurred. But as they zoom in on a crime, everyone else loses their sense of where in the city it's located. They'd like a second window that displays the entire city, with a marker showing where they're zoomed to.

You have been reading the book *Exploring ArcObjects* (part of the ArcGIS software documentation available at www.esri.com/ExploringArcObjects). In the book, you have found a sample subroutine called CreateOverviewWindow that does just what you need. Part of being a good programmer is being a good thief. Stealing this code will save you a lot of programming time.

In this exercise, you will import the CreateOverviewWindow subroutine and call it from a click event.

1 Start ArcMap and open **ex06a.mxd** in the **C:\ArcObjects\Chapter06** folder.

When the map opens, you see Washington, D.C., layers and the Crime Analysis toolbar, which contains several buttons that you will code in this chapter.

Crime Analysis toolbar

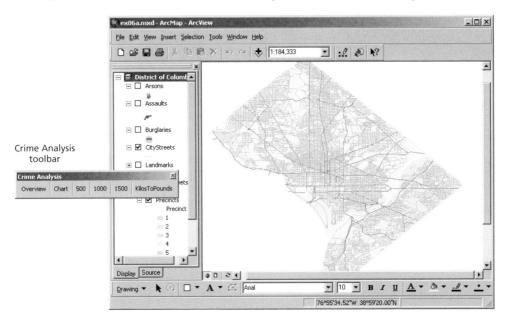

If you have worked through the first five chapters of the book, most of your toolbars may be turned off.

2 If they're off, turn on the Standard, Tools, and Draw toolbars.

3 On the Crime Analysis toolbar, right-click the Overview button and click View Source.

In the ThisDocument code window, you see the Overview click event. This is where you will write the line of code that calls the CreateOverviewWindow subroutine. First, however, you will import the subroutine.

4 In the Project window, right-click **Project (ex06a.mxd)** and click Import File.

The code you are going to import was obtained from the *Exploring ArcObjects* book. Since some people may not have the book, the sample code to create an overview window has been provided with this book's data CD.

5 In the Import File dialog box, click the Files of type drop-down arrow and click All Files. Navigate to **C:\ArcObjects\Data\Samples\ExploringArcObjects** and click **CreateOverviewWindow.txt**. Click Open.

You didn't see anything happen, but a new standard code module has been added to the project. Code modules are the windows in which you write and store procedures. When you import code, a new standard module is automatically created to hold it.

6 In the Project window, under Project (ex06a.mxd), click the plus sign next to the Modules folder to open it.

7 In the Modules folder, double-click Module1 to open it.

You might recognize, in a general way, what the code in CreateOverviewWindow does. The first lines are comments followed by several Dim statements for declaring variables. The next lines set those variables with the Set keyword. (You will learn how to set object variables in chapter 9.) The last lines use the familiar object.property syntax to set properties for a blue outline symbol on the rectangle that shows where the ArcMap display area is zoomed to.

The subroutine uses some ArcObjects code that you won't learn about until chapter 10. It's OK if you don't understand the details. In this situation, all you need to know is that when you tell the subroutine to run, it runs and opens an overview window.

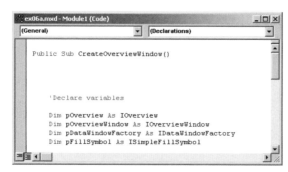

8 In the Properties window, replace the module's name, Module1, with **CrimeAnalysisTasks**. Press Enter.

In the other exercises of this chapter, you will add more subroutines to the CrimeAnalysisTasks module. You could store every subroutine in its own module, but since these are all about the same topic, it's logical to store them together.

The event procedures for your UIControls (like the Overview button's click event) are stored in the ThisDocument code module. This means that procedures in one code module (ThisDocument) will be calling procedures from another code module (CrimeAnalysisTasks). For this to work, the called procedure must be declared Public. CreateOverviewWindow above is already declared Public. Procedures that are declared Private can only be called by other procedures in the same module.

9 Make the project's ThisDocument code window active.

You are ready to code the click event procedure of the Overview UIButton.

10 In the Overview click event of ThisDocument, add the following line of code.

```
Call CreateOverviewWindow
```

The Call statement tells the CreateOverviewWindow subroutine to run.

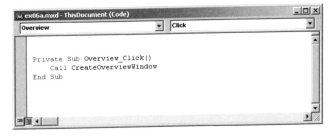

You could have copied all the code in the CreateOverviewWindow subroutine into the click event procedure, dispensing with the Call statement. But by keeping the subroutine and the click event separate, the click event stays uncluttered.

Another benefit of keeping the procedures separate is that you can call CreateOverviewWindow from procedures other than the click event.

11 Close Visual Basic Editor.

There have been several recent burglaries near the Holy Name College. You will test the subroutine by examining the area around the college.

12 Turn the Landmarks and Burglaries layers on.

13 Click the View menu, point to Bookmarks, and click Holy Name College.

Because you're zoomed in, it's hard to tell where in the District the college is located.

14 On the Crime Analysis toolbar, click the Overview button.

You may have to move or resize the Overview window to see ArcMap. In the Overview window, you see a blue box that shows you are zoomed in on a northeast section of the District.

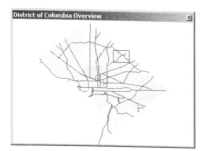

To see crimes in other parts of the city, you can move or resize the blue box in the Overview window. Any changes you make to the blue box affect the ArcMap window.

15 In the Overview window, drag the blue box to the eastern corner of the District, as shown.

The ArcMap display pans to the new area. The extent of the blue box in the Overview window always matches the extent of the ArcMap display. When you make a change to one, the other reacts.

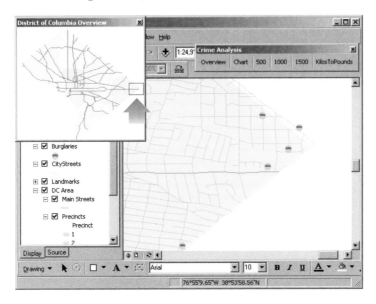

In this exercise, you imported and called a subroutine written by someone else. Programmers do this all the time to make life easier. Sharing code can be as easy as copying it and calling it. Of course, you'll have to alter the code sometimes, but as you become more fluent in VBA and ArcObjects, that will get easier. You'll do more copying and calling in the following exercises.

> ### GET MORE FREE CODE SAMPLES
> Programmers of ArcGIS and other software use the following Web site to share their GIS code:
> ## arcscripts.esri.com

16 Close the overview window.

17 If you want to save your work, click the File menu in ArcMap and click Save As. Navigate to **C:\ArcObjects\Chapter06**. Rename the file **my_ex06a.mxd** and click Save. If you are continuing with the next exercise, leave ArcMap open. Otherwise close it.

Passing values to a subroutine

Sometimes procedures perform their task without any arguments. In the last exercise, you called the CreateOverviewWindow subroutine and it opened an overview window. That subroutine has no arguments. When called, it runs the same way every time.

Arguments provide information to a procedure so that it can run with some variation. Suppose you want to create a subroutine that prints a map at a page size specified by the user (letter-, legal-, C-, D-, or E-size paper). When you code the subroutine, you define an argument with a name and data type in parentheses. It's like declaring a variable, but without the Dim keyword.

```
Public Sub PrintMap (aPageSize As String)

End Sub
```

When you call the PrintMap subroutine, you must enter a value for its argument. This is called passing a value to a subroutine. The code below passes the "Letter" page size to the PrintMap subroutine.

```
Call PrintMap ("Letter")
```

As the PrintMap subroutine runs, the variable declared in its arguments list is set to hold the passed value. Code inside the subroutine can use the variable to respond differently according to which value is passed. For example, your code might use a Case statement to evaluate the variable and run different blocks of code.

```
Public Sub PrintMap (aPageSize As String)
    Select Case aPageSize
        Case "Letter"
            MsgBox "The page size is " & aPageSize
        Case "Legal"
            Some other code runs
    End Select
End Sub
```

When PrintMap is called and passed the Letter value, the code in its Letter Case runs.

Exercise 6b

Up to now, crime analysts have used the ArcMap graph tool to make bar charts. However, to use the graph tool, an analyst must fill out three dialog boxes. They have made a request to speed this process up.

Instead of writing the new code yourself, you go to ESRI's ArcObjects developer help Web site (edn.esri.com). There you do a search and find a sample subroutine called CreateNewChart, which has code to make a chart with one button click. When you get a free sample of code, it probably won't do exactly what you want. This one's flaw is that every chart it produces is called My Chart.

In this exercise, you will import the CreateNewChart subroutine and modify it to accept an argument for the chart title. That way, anyone making a chart can give it the name they want. You'll test the new subroutine by making a chart of arsons per precinct.

1 Start ArcMap and open **ex06b.mxd** in the **C:\ArcObjects\Chapter06** folder.

When the map opens, you see the District layers and the Crime Analysis toolbar.

2 On the Crime Analysis toolbar, right-click the Chart button and click View Source.

In the ThisDocument code module, you see the empty Chart click event procedure. First, you will import the sample code for creating charts, then you will tell it to run from this click event.

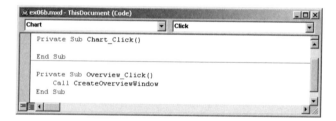

3 In the Project window, right-click **Project (ex06b.mxd)** and click Import file.

The code you are going to import was obtained from the developer help Web site. Since some people may not have an Internet connection, the sample code to make charts has been provided with this book's data CD.

4 In the Import File dialog box, click the Files of type drop-down arrow and click All Files. Navigate to **C:\ArcObjects\Data\Samples\ArcObjectsOnline** and click **CreateNewChart.txt**. Click Open.

A new standard code module is added to the project. To see it, you will open the Modules folder under Project (ex06b.mxd).

5 In the Project window, under Project (ex06b.mxd), click the plus sign next to the Modules folder to open it.

6 In the Modules folder, double-click Module1 to open it.

7 Highlight the entire subroutine from the Public Sub line to the End Sub line. Then right-click the code and click Copy.

You will open the CrimeAnalysisTasks code module so you can paste the code in it.

8 In the Project window, under Project (ex06b.mxd), under Modules, double-click CrimeAnalysisTasks to open it.

You see the sample CreateOverviewWindow subroutine from the last exercise.

9 With the cursor at the top of the module, above the CreateOverviewWindow subroutine, right-click and click Paste.

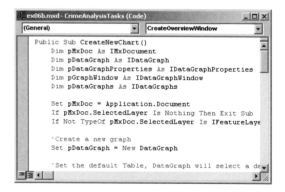

The CreateNewChart code is now in the CrimeAnalysisTasks module. You will edit the subroutine and add the chart title variable to its arguments list, but before you do, you will remove the module that was added to your project when you imported the CreateNewChart sample code.

10 In the Modules folder, under Project (ex06b.mxd), right-click Module1 and click Remove Module1. Click No on the dialog that asks if you want to export the module.

11 In the CrimeAnalysisTasks module, inside the CreateNewChart subroutine's arguments list (between the parentheses), type **strTitle As String**. This declares a string variable for the chart title.

```
Public Sub CreateNewChart(strTitle As String)
```

Next you will find and edit the line of code that sets the chart's Title property. You will replace My Chart with the strTitle variable.

12 Scroll down in the CreateNewChart subroutine and locate the following line.

```
pDataGraphProperties.Title = "My Chart"
```

13 In that line, change "My Chart" to **strTitle**.

```
pDataGraphProperties.Title = strTitle
```

Make sure to use only the variable name and no quotes. Now when any procedure calls CreateNewChart and passes it a text string, the subroutine will store that string in the strTitle variable and use it to set the chart's Title property.

14 Make the project's ThisDocument code window active.

15 In the Chart click event procedure, add the following two lines of code to declare a string variable and use an InputBox to get a chart title from the user.

```
Dim userTitle As String
userTitle = InputBox ("Enter a chart title")
```

16 In the click event, add one more line of code to call the CreateNewChart subroutine and pass it the user's title.

```
Call CreateNewChart(userTitle)
```

The code is now ready to test.

17 Close the Visual Basic Editor window.

18 In the ArcMap table of contents, click the Precincts layer to select it.

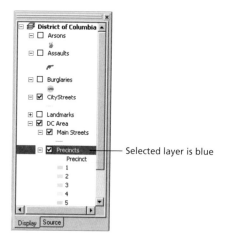

Selected layer is blue

The CreateNewChart code runs on the selected layer.

19 On the Crime Analysis toolbar, click the Chart button.

20 In the dialog box, type **Arsons in 2002**.

21 Click OK.

The CreateNewChart subroutine uses the selected layer's attribute table to make the chart. The chart values, showing the number of arsons, are taken from the first numeric field in the table. The chart labels, identifying the precincts, are taken from the first string field in the table.

The chart shows the number of arson cases by precinct. Precinct 4 (the red bar) has twenty cases and Precinct 6 has none. If it had any, you would see a yellow bar.

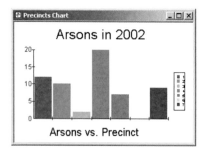

22 Close the chart.

23 If you want to save your work , click the File menu in ArcMap and click Save As. Navigate to **C:\ArcObjects\Chapter06**. Rename the file **my_ex06b.mxd** and click Save. If you are continuing with the next exercise, leave ArcMap open. Otherwise close it.

Making several calls to a single subroutine

Sometimes you design a user interface with two controls that do the same thing. ArcMap, for example, has the Layout View menu choice on the View menu and the Layout View minibutton in the lower left corner of the map display. Both switch the view to Layout View.

Since they do the same thing, you don't have to write the same code twice. You can write a single "Switch to Layout View" subroutine and call it from the event procedures of the two controls.

Switch to
Layout View —

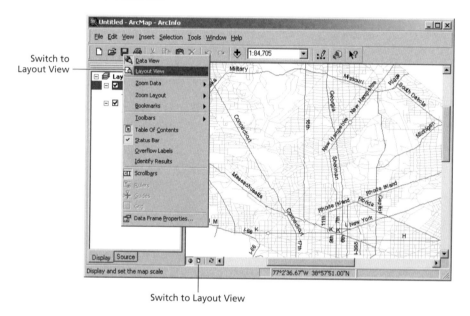

Switch to Layout View

Exercise 6c

Arsonists commonly don't stop with setting just one fire, and, oddly enough, they don't generally go very far afield. So the people who analyze arsons typically start by drawing buffer zones to see if there are similar arsons near each other.

Instead of writing the code yourself, a search of the developer help yields a sample subroutine called BufferFeatures. It buffers each selected feature in the focus map and stores the results as polygons.

In this exercise, you will import the subroutine into Visual Basic Editor and call it from three different buttons. Once again, you will modify the code to accept an argument. This time, each button will pass a different buffer value: 500, 1,000, or 1,500 meters.

1 Start ArcMap and open **ex06c.mxd** in the **C:\ArcObjects\Chapter06** folder.

When the map opens, you see the District layers and the Crime Analysis toolbar. The Arsons layer is turned on.

2 On the Crime Analysis toolbar, right-click the 500 button and click View Source.

In the ThisDocument code module, you see the empty Buffer500 click event procedure. After importing and copying the procedure for drawing buffers, you will call it from this click event.

3 In the Project window, right-click **Project (ex06c.mxd)** and click Import file.

The code you are going to import was obtained from the ArcObjects developer help. Since some people using this book may not have this help system installed on their computer, the sample code to buffer selected features has been provided with this book's data CD.

4 In the Import File dialog box, click the Files of type drop-down arrow and click All Files. Navigate to **C:\ArcObjects\Data\Samples\ArcObjectsDeveloperHelp** and click **BufferFeatures.txt**. Click Open.

A new code module is added to the project. To see it, you will open the Modules folder under Project (ex06c.mxd).

5 In the Project window, under Project (ex06c.mxd), click the plus sign next to the Modules folder to open it.

6 In the Modules folder, double-click Module1 to open it.

7 Highlight the entire subroutine from the Public Sub line to the End Sub line. Then right-click on the code and click Copy.

You will open the CrimeAnalysisTasks code module, so you can paste the code in it.

8 In the Project window, under Project (ex06b.mxd), under Modules, double-click CrimeAnalysisTasks to open it.

You see the subroutines from the previous exercises.

9 With the cursor at the top of the CrimeAnalysisTasks module, above the CreateNewChart subroutine, right-click and click Paste.

The first line of the subroutine has an empty arguments list. You will add an argument to allow the passing of a buffer distance. The 500, 1000, and 1500 buttons on the Crime Analysis toolbar will each call this subroutine and pass it different distance values.

Before you adjust the code to suit your needs, you will remove the module that was added to your project when you imported the BufferFeatures sample code.

10 In the Modules folder, under Project (ex06c.mxd), right-click Module1 and click Remove Module1. Click No on the dialog that asks if you want to export the module.

11 Inside the argument list of the BufferFeatures subroutine (between the parentheses), type **strBufferDistance As String**. This declares a string variable to hold the buffer distance.

```
Public Sub BufferFeatures(strBufferDistance As String)
```

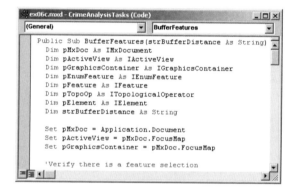

Recall that the chart title argument from the last exercise gave the user more freedom; here, your buffer distance argument will be used to restrict their freedom. Instead of giving the user an input box to enter a distance, your code will pass one of three predefined values.

By the way, it may seem odd to work with distance as a string, but the existing code is written to expect a string. You could change it so that it expects a number, but leaving it as a string is easier and will get the job done just as well.

You have to make one other change. The BufferFeatures subroutine already has two lines of code that declare and set a buffer distance variable. Since you have put your own buffer distance variable in the arguments list, you will find and delete these two (nonadjacent) lines of code that are no longer needed.

12 Near the top of the subroutine, locate the following line of code.

```
Dim strBufferDistance As String
```

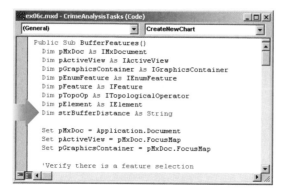

This is the line that declares a buffer distance variable.

13 Delete the line of code.

14 Scroll down in the subroutine and locate the following line of code.

```
strBufferDistance = InputBox("Enter Distance:", "Buffer")
```

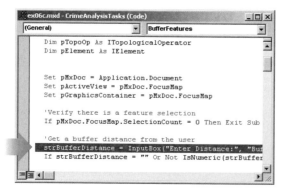

This is the line where the user sets a buffer distance. In your code, the buffer distance will be passed in from other procedures.

15 Delete this line of code, too.

16 Make the project's ThisDocument code window active.

Next you will go to the click events of the three buffer buttons and add a line of code that calls the BufferFeatures subroutine.

17 In the ThisDocument module, add the following line of code to the Buffer500 click event to call the BufferFeatures subroutine and pass it 500 as a text string.

```
Call BufferFeatures("500")
```

18 At the top left of the ThisDocument code window, click the object list drop-down arrow and click Buffer1000. In the Buffer1000 click event, add the following line of code to call the BufferFeatures subroutine and pass it 1000.

```
Call BufferFeatures("1000")
```

19 Click the object list drop-down arrow and click Buffer1500. In the Buffer1500 click event, add the following line of code to call the BufferFeatures subroutine and pass it 1500.

```
Call BufferFeatures("1500")
```

20 Close Visual Basic Editor.

In the past year, many arsons have occurred near Howard University. You will use the buffer buttons to analyze them.

21 In ArcMap, click the View menu, point to Bookmarks, and click Howard University to zoom to it.

You will select the university and buffer it.

22 If your Tools toolbar is turned off, turn it on now.

23 On the Tools toolbar, click the Select Features tool. Then click on Howard University (the long yellow feature) to select it.

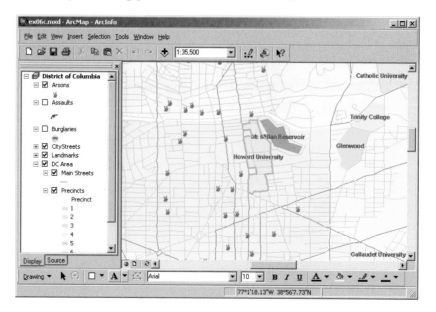

24 On the Crime Analysis toolbar, click the 1500 button. Click the 1000 button. Click the 500 button.

The buffer zones draw in blue. In later chapters, you will write code to control the symbols and colors used for drawing. For now, you will select the three buffer graphics and change their symbology manually.

25 Click the Edit menu and click Select All Elements.

26 Right-click inside any buffer zone and click Properties. In the Common Properties for Selected Elements dialog box, click Change Symbol.

27 Scroll to the bottom of the Symbol Selector and double-click the symbol called Crime Reporting Sector. Click OK on the Common Properties dialog box.

Several arsons have occurred within each buffer zone around the university.

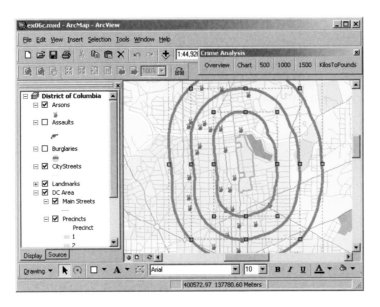

28 If you want to save your work, click the File menu in ArcMap and click Save As. Navigate to **C:\ArcObjects\Chapter06**. Rename the file **my_ex06c.mxd** and click Save. If you are continuing with the next exercise, leave ArcMap open. Otherwise close it.

Returning values with functions

In the previous exercises in this chapter, you called subroutines and passed values to them with arguments. Functions work similarly, except that (as mentioned at the start of the chapter) a function returns a value. To be more specific, a function returns a value to the same line of code that calls it. You'll see how this works in a moment.

A function is like a calculator. You push buttons on a calculator to pass values to it, such as 1 and 2, and to define an operation, such as addition. The calculator then does some arithmetic and returns a new value (with any luck, it will be 3). When you call a function, you pass it a value with a line of code, it carries out some operation on that value, and it passes a new value back.

You have already worked with a couple of VBA functions: MsgBox and InputBox. The line of code below calls the InputBox function. You pass the InputBox function a text string inside its parentheses. The string displays in a dialog box that prompts the user to type in a value. Once a value is entered, it is passed back out of the InputBox function and assigned to the strValue variable.

```
strValue = InputBox("Enter a Parcel Value")
```

This line of code calls the function and passes it a value (to the right of the equals sign), and it receives a new value back from the function and holds it in a variable (to the left of the equals sign).

The code behind VBA's built-in functions is hidden from you. When you make your own function, you have to write the code yourself (or steal it).

Suppose you want to make a function that converts kilometers to miles. The function will receive an input value in kilometers, do some arithmetic, and pass back the equivalent in miles. To code a function, you start with its wrapper lines:

```
Public Function KilometersToMiles (km As Double) As Double

End Function
```

The first line includes the name of the function (KilometersToMiles); an argument declaring a variable and data type for the input value (km As Double); and the As keyword followed by a data type for the returned value (As Double, shown here in boldface).

Although they are of the same data type here, the input value and the returned value do not have to be. Say you were creating a function to calculate someone's age based on the date they were born. The input value might be of the data type Date and the return value might be of the data type Integer.

```
Public Function AgeInYears (age As Date) As Integer

End Function
```

To complete the code for the KilometersToMiles function, you need a conversion formula to change kilometers into miles (1 kilometer equals 0.621371 miles) and you need to pass the new value out of the function and back to the line that called it. You can do all that with one line of code.

To pass a value out of a function, you use the function's name like a variable. The line of code below sets KilometersToMiles equal to the miles equivalent of kilometers.

```
Public Function KilometersToMiles (km As Double) As Double
    KilometersToMiles = km * 0.621371
End Function
```

Setting KilometersToMiles as if it were a variable allows the line of code that calls the function to receive the returned value.

When you call functions you don't use the Call keyword as you do with subroutines. Instead, you typically use two lines of code. The first declares a variable to hold the value that the function will return. The second sets this variable equal to the function name and passes a value into the function (10 in the example below). The second line is the one that actually calls the function. In the example below, the variable's data type is Double because the function's return value is a Double.

```
Dim x As Double
x = KilometersToMiles (10)
```

The following diagram illustrates the process. The function is coded in the ConversionTasks module and called from the ThisDocument module, where the value 10 is passed into it. In the function, this input value is stored in the km variable and converted to 6.21371 miles. Then KilometersToMiles is set equal to 6.21371, which causes the value to be passed out of the function back to the line of code that called the function. There the returned value is set equal to the variable x. When all is said and done, x holds the value 6.21371.

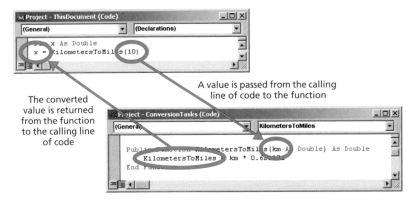

A value is passed from the calling line of code to the function

The converted value is returned from the function to the calling line of code

Declaring and setting a variable isn't the only way to handle a function's returned value. Since a function returns a value, the function can be used anywhere that that value is accepted. For example, message boxes display a value, so you can use the

KilometersToMiles function as a message box's first argument as shown on the line below.

```
MsgBox KilometersToMiles (10)
```

Exercise 6d

During the crime analysis meetings, the group reviews recent drug busts. The amount of drugs confiscated is typically reported in kilograms, and someone always asks, "How much is that in pounds?"

In this exercise, you will make a function to convert kilograms to pounds. You will call the function from a button on the Crime Analysis toolbar. Once the function is written, you will be able to call it from any other procedures that have a use for it.

1 Start ArcMap and open **ex06d.mxd** in the **C:\ArcObjects\Chapter06** folder.

When the map opens, you see the familiar District layers and the Crime Analysis toolbar.

2 Click the Tools menu, point to Macros, and click Visual Basic Editor.

3 In the Project window, expand Project (ex06d.mxd) and expand the Modules folder underneath it. Double-click CrimeAnalysisTasks to open the code module.

In the CrimeAnalysisTasks module, you see the other subroutines that you have worked with in this chapter.

4 With the CrimeAnalysisTasks module active, click the Insert menu and click Procedure.

5 In the Add Procedure dialog box, type **KilogramToPound** as the name. For the type, click Function. Leave the scope set to Public.

6 Click OK.

The new function is added to the bottom of the CrimeAnalysisTasks module. You'll define its arguments and returned value data type in the next two steps.

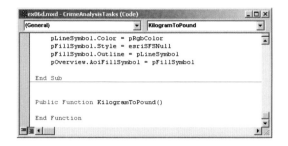

7 In the function's arguments list (between the parentheses), type **dblKilos As Double**.

This defines a variable to hold the value that is passed in to the function and specifies its data type as Double.

```
Public Function KilogramToPound(dblKilos As Double)
```

8 Define the data type of the function's returned value by typing **As Double** after the function's arguments list.

```
Public Function KilogramToPound(dblKilos As Double) As Double
```

Next you will code the conversion formula. There are 2.2046 pounds to a kilogram, so the formula is pounds = kilograms * 2.2046.

9 Inside the function, set the function name equal to the formula to convert kilograms to pounds.

```
KilogramToPound = dblKilos * 2.2046
```

dblKilos holds the kilograms value that is passed into the function. This value is converted into pounds by multiplying it by 2.2046. To return a value from the function to the line of code that called it, you set the function's name equal to a value. Here it's the formula's result value.

The function is now complete. All that remains is to code the KilosToPounds button's click event procedure to call the function and pass it a value.

10 Make the ArcMap application window active. On the Crime Analysis toolbar, right-click KilosToPounds and click View Source.

You will use an InputBox to get a kilogram value from the user. Whatever value the user types into the InputBox is the value that will be passed to the KilogramToPound function. Since this value is not predetermined, you will declare and set a variable for it.

11 Inside the KilosToPounds click event, declare a variable for the user's kilogram value and set it using the InputBox function.

```
Dim userKilos As Double
userKilos = Inputbox ("Enter the number of kilograms")
```

You are ready to call the KilogramToPound function. This includes declaring a variable to hold the function's returned value.

12 Declare a variable for the KilogramToPound function's returned value and set it by calling the KilogramToPound function.

```
Dim userPounds As Double
userPounds = KilogramToPound (userKilos)
```

13 Add a final line of code that uses a message box to report the converted value.

```
MsgBox userKilos & " kilograms is " & _
       userPounds & " pounds"
```

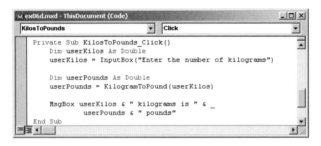

The button and the function are ready to test.

14 Close Visual Basic Editor.

15 On the Crime Analysis toolbar, click the KilosToPounds button.

16 In the input dialog box, type **14** and click OK.

You see a message box that shows the kilogram value compared to the pound value.

17 Click OK on the message.

18 If you want to save your work, click the File menu in ArcMap and click Save As. Navigate to **C:\ArcObjects\Chapter06**. Rename the file **my_ex06d.mxd** and click Save. If you are continuing with the next chapter, leave ArcMap open. Otherwise close it.

Looping your code

Coding a For loop

Coding a Do loop

In VBA programming, tasks can be run repeatedly with looping statements. Loops can be as simple as "Print ten copies of the same map," or as complex as "For each vacant parcel in the city, get its acreage, add that to a running sum, and report the total acreage of vacant land." VBA has two kinds of looping statements: For loops and Do loops. For loops run a given number of times and Do loops run until the value of a logical expression changes.

You are going to make lunch for yourself and two friends. All three of you will have a peanut butter and jelly sandwich. To make the sandwiches, you repeat the following process three times: get two slices of bread, spread peanut butter on one, spread jelly on the other, slap them together, cut them in half, and put them on a plate for serving.

The sandwich loop above is a For loop. You are making one sandwich for each person, and so the loop runs a specified number of times. When everyone has a sandwich, the loop ends (and the eating subroutine begins).

A Do loop evaluates a logical expression and then decides whether to run its block of code. A Do loop will run its code until the expression's true or false status changes.

You are so good at making peanut butter and jelly sandwiches that now it's your job to make them from 8 to 5. When the expression "time > 5:00 P.M." changes from false to true, your sandwich-making stops for the day. So every time you finish a sandwich, you look up at the clock to check the time. If it's not yet five o'clock, you have to make another sandwich. If it's five, you get to stop.

Coding a For loop

The mechanics of a For loop are simple. You set a variable equal to a start value, like 1. Each time your block of code runs, the variable value increments by 1 until it reaches an end value, like 10.

A For loop begins with the For keyword and ends with the Next keyword. The For keyword on the first line is followed by the variable and its start and end values. The block of code between this line and the Next keyword runs repeatedly until the variable exceeds the end value.

```
For variable = StartValue To EndValue
    Block of code here
Next
```

The For loop below would print a map ten times. The first time through the loop, the variable x is set equal to 1 (the start value) and the block of code runs to print the first map. The second time through the loop, x is set equal to 2 and a second map prints. The third time, x changes to 3 and a third map prints. The loop stops running after the tenth map is printed.

```
For x = 1 to 10
    'Code here to print map
    MsgBox "Printing Map " & x
Next
```

It is common to begin with 1 and end with the number of times that you need the loop to run; however, you can use any starting and ending values you like.

If your loop is processing a list of items, you can also skip over items using the Step keyword. Suppose you have a database of 10,000 customers and you only have enough money to mail a coupon to 5 percent of them. Your oldest customer is customer 1 and your newest is customer 10,000. You want the coupons to go to a mixture of new and old customers.

The loop below is designed to go to every twentieth record starting with record 1. For every twentieth record, the customer's address is printed on a mailing label.

```
For x = 1 to 10000 Step 20
    'in the customer table go to record x
    'get address for x
    'print a mailing label
Next
```

Looping statements are also useful for populating lists, like those in combo boxes. The only way to add values to a combo box is with the ComboBox's AddItem method. Suppose you wanted to make a combo box in which the user can choose a number from one to ten.

To put the numbers one through ten in the combo box, you could write ten lines of code with the AddItem method.

```
cboFavoriteNumber.AddItem 1
cboFavoriteNumber.AddItem 2
cboFavoriteNumber.AddItem 3
cboFavoriteNumber.AddItem 4
cboFavoriteNumber.AddItem 5
cboFavoriteNumber.AddItem 6
cboFavoriteNumber.AddItem 7
cboFavoriteNumber.AddItem 8
cboFavoriteNumber.AddItem 9
cboFavoriteNumber.AddItem 10
```

Luckily, there's a faster way—you can write a For loop and reduce those ten lines of code to three. The loop below increments from 1 to 10, adding those values to the combo box's list.

```
For x = 1 to 10
    cboFavoriteNumber.AddItem x
Next
```

Exercise 7a

As a GIS programmer for the U.S. Census Bureau, you are developing an application to help people view county-level population trends. Your application will display population values for each census year from 1930 to 2000. When a user chooses a year, and picks a number of population classes for the legend, a map is created.

The Quit and MakeMap buttons on the dialog box below are already coded; the two combo boxes, however, are not. In this exercise, you will code two For loops to add values to the combo boxes. The box called cboYears will contain a list of the census decades from 1930 to 2000. The box called cboClasses will contain a list of numbers that sets the number of population classes in the legend.

Legend with 3 classes

☐ 𝕊 **Layers**
  ☐ ☑ Counties
    1930
     ☐ 0 - 265804
     ▨ 265804 - 1374410
     ■ 1374410 - 3982123

Legend with 5 classes

☐ 𝕊 **Layers**
  ☐ ☑ Counties
    1930
     ☐ 0 - 71235
     ☐ 71235 - 252312
     ▨ 252312 - 634394
     ▨ 634394 - 1374410
     ■ 1374410 - 3982123

1 Start ArcMap and open **ex07a.mxd** in the **C:\ArcObjects\Chapter07** folder.

When the map opens, you see a layer of U.S. counties. On the Standard toolbar, to the right of the Add Data button, is a button called CensusMaps.

2 Click the CensusMaps button.

On the Census Population Maps dialog box, you see the cboYears and cboClasses combo boxes. If you clicked their drop-down arrows, you would see that both are empty. You will write two loops to fill these combo boxes with values.

3 Click Quit to close the dialog box.

4 Click the Tools menu, point to Macros, and click Visual Basic Editor.

5 In the Project window, under Project (ex07a.mxd), under Forms, double-click frmCensus to open it.

Before writing any code, you will make a property setting for the two combo boxes to help eliminate user typing errors.

6 On the form, click the combo box cboYears to select it. In the Properties window, set its Style property to 2-fmStyleDropDownList.

This option forces the user to pick a decade from the drop-down list. Otherwise, they could type in any value they liked. Since you only have data for the census years between 1930 and 2000, you don't want the user to type in a number (like 1981 or 1920) that will cause an error.

Next, you will set the same Style property for cboClasses to keep users from making maps with fewer than three or more than eight classes.

7 On the form, click the combo box cboClasses to select it. In the Properties window, set its Style property to 2-fmStyleDropDownList.

Now you will write the code for the looping statements.

8 At the top of the Project window, click the View Code button.

The frmCensus code module becomes active. It has code in it for the MakeMap and Quit buttons. The Quit button's click event has code to close the form. The MakeMap button's click event has code to call the RenderMap subroutine.

The RenderMap code uses the selected combo box values to get census data from the Counties layer attribute table and create a legend for it. You'll learn to write this type of code using ArcObjects later in the book.

9 In the object list, click the drop-down arrow and click UserForm (unless it is already selected). In the procedures list, click the drop-down arrow and click Initialize.

Since the initialize event runs just before the form opens to the user, this is where you will add the two loops to fill the combo boxes with years and numbers.

Initialize event

10 In the initialize event, declare the upcoming For loop's variable as integer.

As the loop runs, this variable will change its value for each integer in the range between the start and end value. You'll set up that range in the next step.

```
Dim intClass As Integer
```

11 Add the following For loop. Use 3 and 8 for the start and end values.

A legend with fewer than three classes doesn't convey much information, and a legend with more than eight is hard to interpret. Visually, the shades of a color start to look alike, and conceptually, the distinctions between classes become less meaningful.

```
For intClass = 3 To 8

Next
```

When a user clicks 3 in cboClasses, a legend with three classes will be created, like the one below.

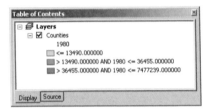

12 Inside the loop, add the following line that uses the AddItem method on the combo box to add the numbers from three to eight.

```
cboClasses.AddItem intClass
```

This line of code will run six times, adding the numbers 3, 4, 5, 6, 7, and 8 as choices to the combo box's drop-down list.

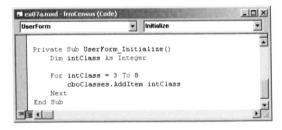

Now you will code a second For loop to add census decades to the cboYears drop-down list.

13 After the code for the first loop, declare an integer variable to hold the For loop numbers for each year.

```
Dim intYear As Integer
```

14 Add the second For loop. Use 1930 and 2000 for the start and end values and make it step every 10 years.

```
For intYear = 1930 To 2000 Step 10

Next
```

15 Inside the loop, add the following line that uses the AddItem method on the combo box to add the year for each decade.

```
cboYears.AddItem intYear
```

This line of code will run eight times because there are eight census years from 1930 to 2000. Each year will be added to the combo box's drop-down list.

So far, when the dialog box opens, the combo boxes are empty until the user clicks the drop-down arrow to make a selection. You will set a property for the combo boxes so they have initial values.

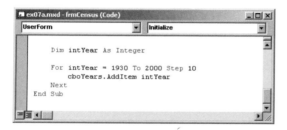

16 After the two loops, add the following two lines of code to set the default values for each combo box.

You will set the number of classes to 3 and the year to 2000. Users can accept these defaults or use the drop-down lists to change them.

```
cboClasses.Value = 3
cboYears.Value = 2000
```

The code is ready to test.

17 Close Visual Basic Editor.

18 Click the CensusMaps button.

The Census Population Maps dialog box opens with initial values in each combo box.

19 Set the year to **1930** and number of classes to **7**.

20 Click MakeMap. Move the dialog box so you can see the map.

Many western counties are yellow, which means they are in the lowest population category for 1930. Next you will look at the populations for 2000.

21 Change the year to **2000** while leaving the number of classes set to 7.

22 Click Make Map.

More western counties are now red and more midwestern counties are yellow. Maybe there was a migration from the midwest to the west. For a more specific analysis, you could zoom in to a specific state and compare population values for different decades.

In the next exercise, you will work with a combo box that does just that. You will populate the combo box with a list of state names and, when the user picks a state, the view will zoom to it.

23 Click Quit.

24 If you want to save your work, click the File menu in ArcMap and click Save As. Navigate to **C:\ArcObjects\Chapter07**. Rename the file **my_ex07a.mxd** and click Save. If you are continuing with the next exercise, leave ArcMap open. Otherwise close it.

Coding a Do loop

Do loops come in two types: While and Until. Do While loops run *while* a logical expression is true; Do Until loops run *until* a logical expression is true (which amounts to running while an expression is false).

Do Loops begin with the Do keyword and end with the Loop keyword. On the first line, the Do keyword is followed by either While or Until and a logical expression. These logical expressions are just like the ones you learned about with If Then statements.

```
Do While Expression
    'Code here runs as long as the expression is true
Loop

Do Until Expression
    'Code here runs as long as the expression is false
Loop
```

The Do While loop below uses a vbYesNo message box in its expression. MsgBox is a VBA function; when used with the vbYesNo argument, it offers a choice of some kind and presents the user with two buttons. Depending on which button the user clicks, the function returns the value vbYes or vbNo.

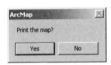

The following expression checks to see if the value returned by the MsgBox function is equal to vbYes. If it is, the expression is true and the loop's block of code runs.

```
Do While Msgbox ("Print the map?", vbYesNo) = vbYes
    'Code here to print map
    MsgBox "Printing Map"
Loop
```

In the previous exercise, you created For loops to add lists of numbers to combo boxes. For loops are good at building lists of this kind, but what if you want to build a list of words, such as a variety of color choices a user can pick for a legend?

To add color names to a combo box, you would have to use the AddItem method for each color. So to put five colors in a drop-down list, you would use AddItem five times.

```
cboColor.AddItem "Red"
cboColor.AddItem "Green"
cboColor.AddItem "Blue"
cboColor.AddItem "Gray"
cboColor.AddItem "Purple"
```

With five colors, this isn't a problem. But what if you wanted to build a combo box list of one hundred colors? You could type in one hundred AddItem lines of code, but that would take a while.

A better way is to access a text file that has all the color names in it. You could then write a Do loop to read the text file, get the names from it, and add them to a combo box.

The process for reading values from a text file goes like this:

```
open the file
    check to see if there is a line in the file
    read the line, do something with it
    repeat these two steps until the end of the file
close the file
```

Let's analyze this process in detail. To open a file, you use VBA's Open function, which has five arguments:

```
Open "c:\names.txt" For Input As #1
```

The first argument is a string with a path to the file. The second argument is the For keyword. The third argument is another keyword: either Input, Output, or Append. (Input is used to read a file, Output to write values to a new file, and Append to add lines to an existing file.)

The fourth argument is the As keyword. The fifth argument (#1) is a file number you assign. In any given session, your code may read information from many different files, so you need a simple way to tell them apart.

When your loop is done reading values, you use the Close function to close the file. The Close function has an argument to specify which file to close.

```
Close #1
```

How do you know when your loop has read all the values? You use the VBA function called End of File (EOF). EOF returns True when the end of a file is reached and False as long as there are more lines to read. Because it returns True or False, it can be used as the Do loop's expression.

In the example below, the Do Until loop applies the EOF function to file number 1. The loop runs as long as EOF is false. When the last line has been read, the EOF function returns True and the loop ends.

```
Do Until EOF(1)
    'Read lines from the file
Loop
```

Inside the loop, you use the Input statement to read the file. Input has two arguments, the file number and a variable. As each line is read, its contents are put into the variable (strName, in this example).

```
Input #1 strName
```

Suppose the line being read from the text file consists of the word "Blue." After the line of code above runs, strName will hold the value "Blue." The variable holds whatever text string it finds in each successive line of the file. In a list of colors, it will hold values like "Blue," "Orange," and "Maroon." In a list of businesses, it would hold values like "ESRI, 380 New York St, Redlands, 92373."

With each iteration of the loop, Input goes to the next line in the text file, gets the value from that line, and sets the variable with that value. When the end of the file is reached, the process stops.

Once you have read the value into the variable, you can do whatever you want with it. In the following example, you use it to add a choice to a combo box.

The code opens a file called names.txt and loops through it. Each line of the text file is successively stored in the variable strName. The AddItem method then takes the variable value and adds it as an item to the combo box. When the end of the file is reached, the loop ends and the text file is closed.

```
Open "c:\names.txt" For Input As #1

    Do Until EOF(1)
        Input #1 strName
        cboName.Additem strName
    Loop

Close #1
```

Exercise 7b

In this exercise you will write a Do loop to read the fifty U.S. state names from a text file and put them in a combo box's drop-down list.

1 Start ArcMap and open **ex07b.mxd** in the **C:\ArcObjects\Chapter07** folder.

When the map opens, you see a layer of U.S. counties with each state outlined.

2 Click the CensusMaps button. Move the dialog box, so you can see it and the map at the same time.

Most of the code behind the dialog box has already been written. When the user picks a state from the States combo box, the view will zoom in on that state.

This combo box itself, however, has not been coded, so its drop-down list is empty. Its drop-down style hasn't been set either, so you can still type into it. Before adding items to the combo box, or setting its Style property, you will make sure the zoom functionality is working.

3 In the Choose a state box, click the drop-down arrow to confirm that it is empty, then type **Utah**.

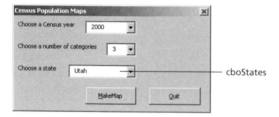

cboStates

4 Click MakeMap.

MakeMap gets Utah from the combo box, zooms to it, and draws its outline.

5 Click Quit to close the dialog box.

6 Outside of ArcMap use a file browser, like Windows Explorer, to navigate to **C:\ArcObjects\Data\USA**. Double-click **StateNames.txt** to open it.

This is the text file your Do loop will read. Each line in the text file contains a single state name.

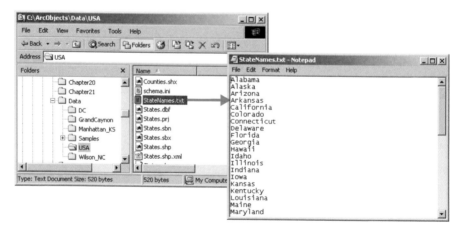

7 Close the text file and file browser windows.

Before writing the Do loop, you'll set the combo box's Style property so that users can't type in the box.

8 In ArcMap, click the Tools menu, point to Macros, and click Visual Basic Editor.

9 In the Project window, under Project (ex07b.mxd), under Forms, double-click frmCensus to open it.

The combo box for state names is called cboStates.

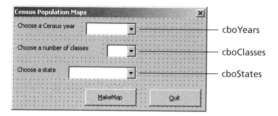

10 Click the cboStates combo box to select it. In the Properties window, set its Style property to 2-fmStyleDropDownList.

11 At the top of the Project window, click the View Code button.

The frmCensus code module opens and you see the code from the previous exercise. The initialize event contains your two looping statements for the year and class combo boxes. Now you'll add a third loop to build the states combo box drop-down list.

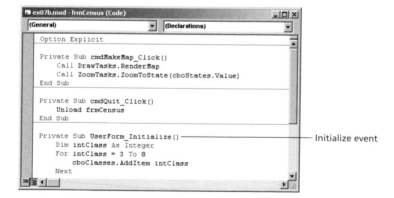

12 At the bottom of the UserForm_Initialize event, just before the End Sub line, add the following two lines of code to declare and set a string variable to hold the text file's path. (If you have loaded your data in a different location, you will have to alter the path below.)

This is the full path to where the text file is located and it includes the name of the text file and its .txt extension. Quotation marks are put around the path to indicate that it is a string.

```
Dim strFile As String
strFile = "c:\arcobjects\data\usa\statenames.txt"
```

13 Add a line of code to declare a string variable to hold the state names.

This variable will hold one state name at a time. Its value will change with each run of the loop.

```
Dim strStateName As String
```

14 Add a line of code to open the text file strFile. Use the Input option and #1 as the file's identification number.

```
Open strFile For Input As #1
```

15 Add a Do Until loop that continues until EOF is true.

The EOF function uses the file number as its argument; in the previous step, you designated the file number as 1.

```
Do Until EOF(1)

Loop
```

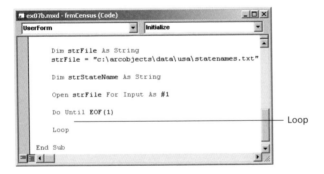

ENDLESS LOOPS

An endless loop means that for some reason the logic behind your loop has no way out. Consider the following loop. The logical expression will never be true, because five will never equal ten. Since this loop's expression is always false, it will run forever.

```
Do Until 5 = 10
Loop
```

The question is, if you accidentally put yourself into an endless loop, how do you escape? One way out is to simultaneously press the Ctrl and Break keys on your keyboard. This operation is called Control–Break.

16 Inside the loop add the following line of code.

```
Input #1, strStateName
```

As the loop runs, this line of code processes each line in the state names text file. The first time through, the strStateName variable is set to hold the first name in the file (Alabama). The second time, the variable is reset to hold the second name (Alaska). The loop ends when the EOF function returns true (there are no more lines to read).

17 Add one more line of code inside the loop that uses the AddItem method to add the statement to the combo box.

```
cboStates.AddItem strStateName
```

18 Outside the loop, after the Loop keyword, add a line of code to close the file.

```
Close #1
```

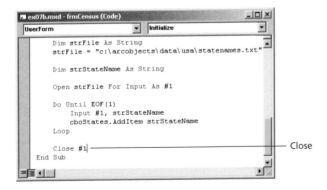

19 Add one last line of code to set the combo box's default value to Alabama.

```
cboStates.Value = "Alabama"
```

20 Close Visual Basic Editor.

21 Click CensusMaps. Move the dialog box so you can see the map.

22 Set the year to **1930**, the number of classes to **3**, and the state to **Nevada**.

23 Click MakeMap.

All Nevada counties are in the lowest population category (yellow), except Washoe county in the northwest. It has a high population because Reno, Sparks, and Lake Tahoe are located there. In 1930, a lot of people were there because of the mining industry.

Next you'll compare the 1930 and 2000 populations.

24 Set the year to **2000**. Don't change the state name or number of categories.

25 Click MakeMap.

Three counties are in the highest third of all U.S. counties (red) and several counties fall in the middle third (orange). Nevada's population is on the rise.

26 Click Quit.

27 If you want to save your work, click the File menu in ArcMap and click Save As. Navigate to **C:\ArcObjects\Chapter07**. Rename the file **my_ex07b.mxd** and click Save. If you are continuing with the next chapter, leave ArcMap open. Otherwise close it.

Fixing bugs

Using the debug tools

Working through the exercises in this book, you have very likely gotten a few error messages. It was probably easy enough to fix your mistakes by comparing what you typed with what was written in the book. As you begin to write your own code, fixing mistakes may not be so simple. However, VBA comes with a Debug toolbar that has several buttons to help you find and fix bugs.

In VBA there are three types of errors: Compile, Run-time, and Logic. In this chapter you will learn what they are, why they happen, and how to find and fix them.

Before your VBA code runs, it is automatically translated into machine language (a language consisting entirely of zeroes and ones, which is all your computer really understands). This translation is called *compiling*. When there is a problem in the translation, you get a compile error and the code does not run. Compile errors occur because you make mistakes with VBA syntax—forgetting to include the Next keyword in a For loop, for instance, or typing a method without the name of an object and a dot, as in the example below.

```
Show
```

If you run code with syntax mistakes, you get a compile error message.

Besides giving you the error message, VBA highlights the offending procedure in yellow and the error itself in blue.

In contrast to compile errors, run-time errors occur after the code has successfully compiled and is being run. A run-time error means that your VBA code is syntactically correct, but that the instructions themselves are impossible to carry out.

The following line of code compiles, because there are two numbers on either side of the division symbol:

```
Acres = 40000 / 0
```

When the line of code runs, however, it causes a run-time error. Division by zero is acceptable syntax, but illegal math.

The following line also compiles, but causes a run-time error:

```
Acres = "SquareFeet" / 43560
```

Here, the type mismatch error appears, because a text string can't be divided by a number.

VBA can't miss compile and run-time errors, and it kindly alerts you to them with error messages. Unfortunately, VBA cannot detect the third type of error, which is a flaw in your program logic. Code that has a logic error compiles and runs, but produces incorrect results—or at least not the results you were aiming for. It's sort of like telling your dog to fetch a ball when you really wanted it to fetch a stick.

Earlier, you saw that trying to divide by zero produces a run-time error. Not all math mistakes will stop your code from running. The line of code below attempts to convert a parcel's square footage to acres. The code compiles and runs, but has a logic error that produces the wrong result every time.

```
aParcelsAcres = aParcelsSquareFeet / 4356
```

The correct formula for acreage is square feet divided by 43560, not 4356. This is an example of a logic error. The syntax is good, the math is legal, but the result is incorrect.

The looping statement below tries to loop until 10 equals 100. Since a Do Until loop runs until the logical expression is true, this loop will run forever.

```
Do Until 10 = 100

    MsgBox x

Loop
```

Simple errors in logic can often be caused by typing mistakes. In chapter 2, you learned that Option Explicit can find typographical errors in variable names. Unfortunately, Option Explicit won't help you with the logic errors above, since no variables are involved.

Logic errors are pesky. The only way to catch them is to test your code with several different inputs for which you already know the correct output. Then have someone else test your code with input values and results that they know.

Using the debug tools

When you get a compile or run-time error, or when a weird result suggests a logic error, you use VBA's Debug toolbar to find and fix your mistake. Sometimes, after reading an error message, you know exactly where the problem is and how to fix it. When you don't, the Step Into button (the Step button, for short) is a good place to start the debugging process.

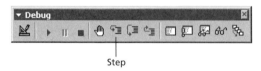

Step

The Step button runs one line of code at a time, while highlighting the line that is about to be run. Suppose you have a ten-line procedure that results in a run-time error. As you step line by line through the code, you will eventually hit the bad line that generates the error message, so you know right where the problem is. Clicking the Run Sub/ UserForm button (the Run button, for short) at any time will run the rest of the code.

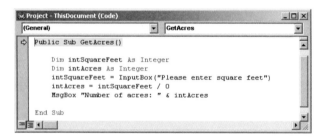

```
Public Sub GetAcres()

    Dim intSquareFeet As Integer
    Dim intAcres As Integer
    intSquareFeet = InputBox("Please enter square feet")
    intAcres = intSquareFeet / 0
    MsgBox "Number of acres: " & intAcres

End Sub
```

But what if you have a procedure with more than a hundred lines? Stepping through each line could take a long time. You can speed things up by using breakpoints, which pause your code at a specific line. If you suspect an error on or after line 50, you can put a breakpoint at line 49. When you click the Run button, all the code runs up to that breakpoint. Then the code pauses and line 49 is highlighted. From there, you could use Step to run one line at a time until you see the error.

You add a breakpoint by highlighting a line and clicking the Toggle Breakpoint button on the Debug toolbar.

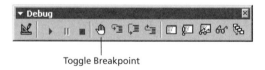

Toggle Breakpoint

You can also add a breakpoint by clicking in the code window's margin. Either way, a red circle marks the spot and the line of code turns red.

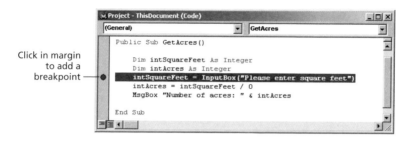

Click in margin to add a breakpoint

When trying to locate an error where a looping statement is involved, you use a combination of stepping, running, and breakpoints. The loop below runs from 0 to 1000. Stepping through it line by line would take about two thousand clicks. (For each iteration of the loop, you would have to click once to enter it and again to finish it.)

Putting breakpoints before and after a loop avoids this problem. Clicking Run on the code below runs it to the first breakpoint. Clicking Run again runs the code to the second breakpoint. Clicking Run a third time runs the code to completion.

First breakpoint

Second breakpoint

At a breakpoint, you can either click Step to run the present line and highlight the next line or click Run to run all the code up to the next breakpoint.

Sometimes, seeing a line of code and an error message isn't enough to determine the cause of the error. While your code is paused on a line, you can view the contents of a variable by hovering the mouse cursor over the variable.

Below, as the mouse is hovered over the intSquareFeet variable in the paused yellow line, a tip appears with the variable's current value of 32000. You can hover over any variable in the procedure, not just those in the paused line of code. If the variable's contents don't match what you expect, you have found the potential cause of an error.

Exercise 8

After a two-week vacation, you return to your GIS programming job at the U.S. Census Bureau to find that a coworker has tried to customize your Census Population Maps dialog box. Unfortunately, many mistakes were made and the dialog box no longer works.

In this exercise, you are going to find and fix the compile, run-time, and logic errors introduced by your coworker.

1 Start ArcMap and open **ex08a.mxd** in the **C:\ArcObjects\Chapter08** folder.

When the map opens, you see U.S. counties and the CensusMaps button on the ArcMap Standard toolbar.

This exercise assumes you are using Visual Basic Editor's default error-trapping option, so you will confirm your setting.

2 Click the Tools menu, point to Macros, and click Visual Basic Editor.

3 In Visual Basic Editor, click the Tools menu and click Options. On the Options dialog, click the General tab. Make sure the Error Trapping option is set to Break on Unhandled Errors.

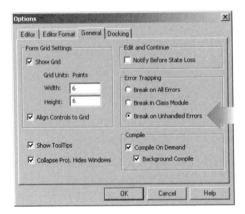

4 Click OK. Then close Visual Basic Editor.

5 Click CensusMaps.

Before your code could compile, a VBA syntax error was located. When a compile error is located, an error message appears, the code stops running, and Visual Basic Editor opens.

6 Click OK on the error message.

The error is highlighted in blue and its procedure in yellow. For some reason, the Show method is on a line by itself without a corresponding form. What can this mean? Is the line supposed to display a form? Was the word "Show" accidentally typed in, or is it a remnant of a previous line of code?

Some comment lines here might have helped answer these questions. It doesn't look as if this Show method has any purpose, but for now, you'll turn it into a comment instead of deleting it.

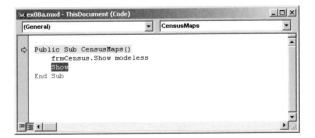

7 Comment out Show by putting an apostrophe in front of it.

```
'Show
```

The commented line turns green. Normally, you use comments to make notes to yourself and other programmers. Comments are also useful if you want to leave in optional lines of code. For example, you might have three blocks of code that do the same task in different ways. You can keep all three, and comment out two of them. Only the uncommented block will run.

Now that you've fixed the compile error, you'll run the code again.

8 Close Visual Basic Editor.

9 Click OK to stop the debugger.

10 Click OK on the User Interrupt message.

Your code was trying to compile and run. You are interrupting that process so you can return to ArcMap and retest the CensusMaps button.

11 Click the CensusMaps button.

Another error appears; this time, it's a run-time error. The message indicates an error in the Project.ThisDocument.CensusMaps procedure.

The type mismatch error means your code uses a data type that VBA is not expecting. For example, you may be using a number where VBA needs a string.

VBA usually highlights the line of code that contains the error. Some errors, however, span more than one line of code and more than one procedure. When that happens, all VBA can do is show you the procedure where it had to stop running.

12 Click OK on the Type mismatch error.

13 Right-click the CensusMaps button and click View Source.

Visual Basic Editor opens to the CensusMaps macro. To help find the error, you will turn on the Debug toolbar and step through the code.

14 In Visual Basic Editor, click the View menu, point to Toolbars, and make sure the Debug toolbar is checked to open it.

15 In the ThisDocument code window, click your cursor inside the CensusMaps procedure. Then, on the Debug toolbar, click Step Into.

Step Into

The first line of code in the macro turns yellow. This line will run as soon as you click Step again.

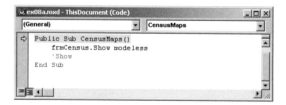

16 Click Step.

The procedure's first line runs. The second line is yellow and ready to run.

17 Click Step.

After you click Step, a form's code window opens and the first line of code in its initialize event procedure turns yellow. Your stepping has taken you into a second procedure in a second code window. The Show method opens a form, but before the form opens, its UserForm_Initialize event must run. So the first line of the initialize event is now yellow and ready to run with the next step.

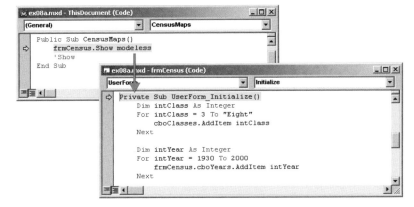

You now see that this error is a bit complicated. For some reason, the Show method was unable to run to completion to open the form. That's why the error message gave the error location as Project.ThisDocument.CensusMaps. However, there isn't anything wrong with the Show line of code. The bad line is actually inside the form's initialize event.

When an error spans multiple procedures, you can see a list of them by clicking the Call Stack button. Call Stack lists the procedures that you are currently stepping through. The initialize event procedure (the current procedure) is at the top of the list.

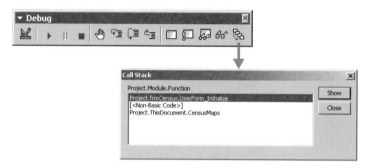

18 Click Step.

The first line of the initialize procedure runs and the first line in the For loop is highlighted. (Dim and other declaration statements are skipped.)

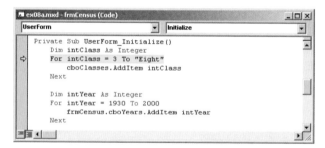

19 Click Step.

A type mismatch appears because the code tries to use a string where an integer is expected. The StartValue is 3, but the EndValue is "Eight". The EndValue should be a number, not a string.

For a detailed description of this error, you can click the error message's Help button. The type mismatch error above is error number 13. Common errors like this have an identification number that you can look up. You can see a list of all errors by searching the Microsoft Visual Basic online help for Trappable Errors.

20 Click OK.

21 Change "Eight" to **8**.

You have fixed the run-time error. The code is now ready to test again.

22 On the Debug toolbar, click Run. (This Run button works just the same as the one on the Standard toolbar.)

Run

ArcMap and the Census Population Maps dialog box appear. If you click MakeMap, it will make a map. However, some people have complained that they get an error message when they pick certain years.

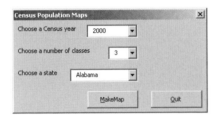

23 Click the Choose a Census year drop-down arrow to see the years that can be selected.

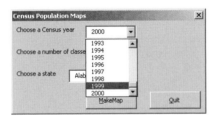

In your original code, only the decade years were added to the drop-down list, but now every year between 1930 and 2000 is there. The Counties layer attribute table, however, only has data for each decade. Having all those extra years in the drop-down list is a logic error—a flaw in the program design—but it leads to a run-time error. If the user picks a nondecade year, there is no corresponding data to display. (It would be a pure logic error if *incorrect* data was shown and no error message was generated.)

24 Click Quit.

Next, you will add a breakpoint and step through the code to find the logic error.

25 Make sure the code module for frmCensus is active.

26 In the UserForm_Initialize event, add a breakpoint by clicking in the margin, to the left of the first line of the Years loop.

The line turns red.

Now you will run all the code up to that line.

27 On the Debug toolbar, click Run.

Your code runs to the breakpoint. The code pauses and that line turns yellow. While it's paused you can run, step, or view variable values.

28 Click Step. Then hover your mouse over the intYear variable.

The variable contains 1930, the first value in the loop.

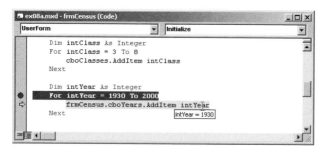

Everything looks normal so far. You'll run the loop again.

29 Click Step twice to run through the loop again. Hover your mouse over the intYear variable.

You expect to see 1940, but you see 1931. A logic error is adding every year to the cboYears combo box. The loop is running as told, but it's not incrementing by 10, because the Step keyword and 10 are missing from the loop's first line (the red line).

Step 10 is missing here

30 Add Step 10 at the end of the first line of the For loop.

```
For intYear = 1930 To 2000 Step 10
```

31 In the margin, click the red circle to remove the breakpoint.

32 On the Debug toolbar, click the Reset button.

The code is reset back to its first line.

Reset

33 Click the Run button.

The dialog box appears.

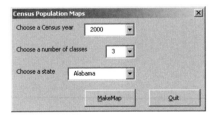

34 Click the Choose a Census year drop-down arrow and click 1950.

Now, only the decade years appear as choices.

35 Set the number of classes to 5 and the state to Texas. Click MakeMap. Close the VB Editor window to see the map.

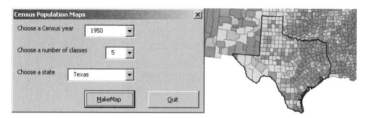

You have fixed the errors introduced by your coworker while you were on vacation. Tactfully, you leave a copy of *Getting to Know ArcObjects* on his desk.

36 Click Quit.

The suggestions below can minimize errors in your code.

- Turn on Option Explicit to identify typing mistakes in variable names.
- Run the code and decipher any error messages.
- Use the Debug toolbar to isolate bad lines of code.
- Test for logic errors by entering values for which you already know the correct results.
- Try all combinations of options on dialog box controls.
- Have someone else test the logic and try all the controls.

37 If you want to save your work, click the File menu in ArcMap and click Save As. Navigate to **C:\ArcObjects\Chapter08**. Rename the file **my_ex08a.mxd** and click Save. If you are continuing with the next chapter, leave ArcMap open. Otherwise close it.

Making your own objects

Creating classes
Creating objects

At an early age children inevitably ask, "Where do babies come from?" In VBA years, you are at that age and ready to ask a similar question, namely, "Where do objects come from?" Fortunately, it's less awkward to answer the second question than the first. Objects come from classes.

You can make a new object with two lines of code. The ones below make a new dog object.

```
Dim d1 As Dog
Set d1 = New Dog
```

Later in the chapter, we'll talk about this code in more detail; right now, the point is that you can make a dog object because somebody somewhere has already made a Dog class. Your dog object comes out of that class. The person who made the class defined what dogs are and what they do. They may have decided that dogs have a Name property and a Bark method. That allows you to write code like:

```
d1.Name = "Sparky"
d1.Bark
```

A class is like a blueprint or template—you can make as many objects from it as you like. Every new object that comes from the Dog class has the same collection of properties and runs the same methods. The code below makes a second dog, sets its name, and makes it bark.

```
Dim d2 As Dog
Set d2 = New Dog
d2.Name = "Rex"
d2.Bark
```

Maybe you're thinking that if objects come from classes, that just raises another question: where do classes come from? Programmers make them, of course—but how?

A class is basically a container full of properties and methods. The container is a code module and the properties and methods are procedures.

You already know about storing procedures in code modules. In chapter 4, you wrote procedures for a Tax Calculator form and stored them in a module called a form module. In chapter 6, you wrote procedures for a CrimeAnalysis toolbar and stored them in a module called a standard module. It's the same thing with classes. You declare variables and write procedures that become the properties and methods of the class, and you store them in a module called a class module.

The following graphic shows the code module for the Dog class, along with its properties and methods:

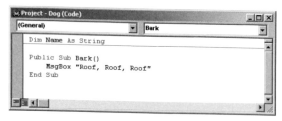

To create a property, you declare a variable and specify a data type. The first line of code in the example creates a Name property that accepts any text string as a value.

```
Dim Name As String
```

To create a method, you write a subroutine or function. The next lines in the example create a Bark method.

```
Public Sub Bark()
    MsgBox "Roof, Roof, Roof"
End Sub
```

Running this method displays a message box.

Every object you make accesses the code in the class module it comes from. When you create a d1 dog and set its Name property to "Sparky," VBA looks at the Name property code in the Dog class module and makes sure that a string is expected. When you tell the dog to bark, your code calls the Bark() subroutine in the Dog class module and runs it.

You can think of object-oriented programming as having two tiers: a lower tier of creating and using objects from existing classes, and an upper tier of creating the classes themselves by writing the code for properties and methods in a class module. Programmers refer to these tiers as client and server programming. A class is called a server since it provides services to clients. When you make a class, you are doing server-side programming. When you write code to create an object from a class and use its properties and methods, you are doing client-side programming. Your code is like a client receiving services.

This book is mainly about client-side programming with ArcObjects servers. However, knowing a bit about server-side programming can only help make you a better client-side programmer.

chapter

9
10
11
12

Creating classes

Microsoft programmers have already created classes for all the VBA objects like forms and controls. ESRI programmers have already created classes for all the GIS objects like maps and layers. You may wonder if there's anything left for you. Luckily, there are plenty of classes to be created that are specific to the problems you must solve in your work.

Say you work for a city and you are building an application to help manage tax assessment and collection. It might be useful to model the city's land parcels in ArcMap. Neither VBA nor ArcGIS has a parcel object. Sure, you could make a Parcel layer from a Parcel Feature Class in a geodatabase and query its attributes. But still, there is no programmable object called a parcel. So you couldn't write a line of code like the one below to display a parcel's value:

```
MsgBox myParcel.Value
```

And you couldn't write a line to display a parcel's zoning code:

```
MsgBox myParcel.Zoning
```

That may give you a reason to create a parcel class of your own. Before you do the programming, however, you need to do some planning. What properties should parcel objects have and what methods should they run?

Say you decide that parcel objects should have a Value property. You want this property to have the Currency data type (since the parcel value is monetary) and you want to be able to get or set it. You also decide that parcels should have a Zoning property. This property will have a String data type (to hold values like "residential" or "commercial") that you also want to be able to get or set. Finally, you want a method that calculates the tax on a parcel.

At first, you might sketch your ideas on paper, but sooner or later it's a good idea to represent them with Unified Modeling Language (UML). UML is a diagramming technique that programmers use to draw classes, properties, and methods with standard symbols. For example, classes (like Parcel) are drawn as rectangles, properties (like Value and Zoning) appear next to barbells, and methods (like CalculateTax) appear next to arrows. These UML pictures are called object model diagrams.

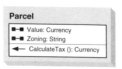

In the diagram, data types are listed to the right of properties and methods. For example, you see Zoning: String which means that the Zoning property is stored as a string.

There are several software packages that draw UML symbols. If you code applications with the participation of other people—especially other programmers—it's probably worthwhile to buy one of them. Whether you use UML or not, it's essential to diagram your class before you begin coding it.

The diagrams are most useful after your coding work is done. People who are new to your classes can look at your diagrams to understand how they work. In the following chapters, you will look at ArcObjects diagrams to learn how those classes work.

In VBA, you create a new class by making a new class module. You can store the module in any of the three projects: the current map document, the normal template, or a base template. If you create the class in the current map document, like the Parcel class module below, client-side programmers will only be able to create parcel objects in that map document. If you want the class to be available to any map document, you should store the class module in the Normal project.

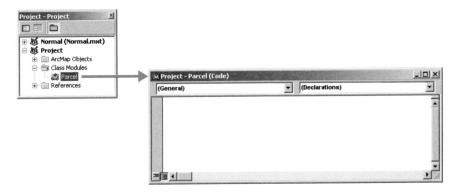

To make a property for a class, you simply declare a variable with a data type. You've decided that the Parcel class will have two properties. You create them by by declaring variables in the class module.

Unlike the variables you've worked with so far, these variables are declared outside a procedure and with the Public keyword, instead of inside a procedure with the Dim keyword. Variables declared with Dim can only be used in the procedure they are declared in. The Value and Zoning variables, however, do not belong to a specific procedure. Declaring them outside a procedure and with Public allows them to be used by any procedure in any code module (as long as the procedure is in the current map document, since that's where you are storing the class).

Properties can also be created by the more advanced method of writing property procedures. (In chapter 2, you read that there are four types of procedures: event, subroutine, function, and property.) With property procedures, you can write more detailed code to control the getting and setting of properties. For example, you might write a procedure that retrieves property values from a database table. While this book does not go into the subject of property procedures, you can learn more about them by searching the Visual Basic online help for the Property keyword.

To make a method for a class, you code a subroutine or a function, depending on whether the method returns a value. The CalculateTax method will return a tax amount, so it would be coded as a function. (The code would be similar to the code you wrote in chapters 4 and 5 to calculate taxes. The difference is that instead of putting it in a click event procedure, which requires user interaction to run, you would make it one of the object's methods, so it could be run behind the scenes.)

The CalculateTax function

On the UML diagram of the Parcel class, the data type returned by the method is listed after its name. CalculateTax(): Currency translates to "The CalculateTax method returns the Currency data type." So you add As Currency to the function's code as shown below. The method has no arguments, so its parentheses are empty.

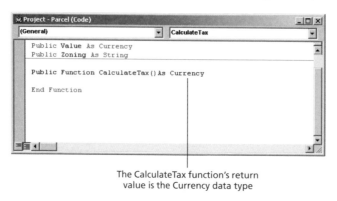

The CalculateTax function's return value is the Currency data type

Classes are code modules, properties are variables (or property procedures), and methods are subroutine or function procedures. The objects that come out of the class may be simple or complex, depending on the code you write for each property and method. Client-side programmers never know the difference, because they get all of that code to run by using the object.property and object.method syntax.

Exercise 9a

As a programmer for a wildlife conservation project, you work with biologists who observe elephants. Currently, the scientists use notebooks to keep track of their elephants. They also record each elephant's trumpeting sound in a .wav file. Your job is to create an ArcMap environment for their work. Since elephant objects don't exist in VBA or ArcGIS, you will program a new elephant class.

The biologists want to store and retrieve elephant names and ages. They also want to be able to hear an elephant's trumpet sound. You can use that information to make a UML diagram of the Elephant class.

In this exercise, you will do the server-side programming to make an Elephant class based on the diagram. Your elephant class will consist of two variables and a procedure that plays a sound file. In the next exercise, you will do the client-side programming and write code to create elephant objects out of the class and use their properties and methods.

1 Start ArcMap and open **ex09a.mxd** in the **C:\ArcObjects\Chapter09** folder.

The map is empty because the Elephant class you are about to create won't be linked to any geographic location. In later chapters, you will learn how to create geographic features like points, lines, and polygons, and assign their geographic coordinates and attribute values.

2 Click the Tools menu, point to Macros, and click Visual Basic Editor.

3 In the Project window, right-click Project (ex09a.mxd), point to Insert, and click Class Module.

Class1 opens and looks just like any other code module. This one will store your Elephant class and the code for its properties and methods.

4 In the Properties window, for the Name property, replace Class1 with **Elephant**. Press Enter.

The Elephant class is created. Now you will add its properties.

5 In the Elephant class module, add the following two lines of code to declare age and name variables.

```
Public Age As Integer
Public Name As String
```

Notice that you do not write code to set these variables. That's because they will be set by client-side programmers (including you) who make elephant objects out of your Elephant class. For example, after creating a new elephant, a client-side programmer might write

```
myElephant.Name = "Benny"
```

or

```
myElephant.Age = 10
```

Now that the properties are done, you will code the Trumpet method.

6 With the Elephant code module active, click the Insert menu and click Procedure.

7 In the Add Procedure dialog box, type **Trumpet** as the name. Make sure the Type is set to Sub and the Scope to Public.

If the Trumpet method were going to return a value, like a decibel level, you would make it a function. Since it just plays a sound file, you'll make it a subroutine.

8   Click OK.

9   Inside the Trumpet subroutine, add the following code. (If you have installed the data at a different location, you will need to modify the path accordingly.)

```
sndPlaySound _
    "C:\ArcObjects\Data\elephant.wav", _
    SND_ASYNC
```

This code is a bit different from anything you've seen before. The sndPlaySound function does not come from either VBA or ArcObjects, but is a Microsoft Windows Application Programming Interface (API) function. API functions carry out operating system operations like playing sound files, finding out who is logged on to a computer, or retrieving the path to the Temp folder.

To call an API function in VBA, you first have to declare it in another code module. (You did this in chapter 6 when you declared the KilogramToPound function in one module and called it from another.) In this exercise, the function has already been declared for you in the standard module called PlaySounds.

The sndPlaySound function has two arguments: the first is a path to a .wav sound file, and the second is a constant, which can be either SND_SYNC or SND_ASYNC. The SND_SYNC option pauses the code until the sound file has finished playing. The SND_ASYNC option lets more code run while the sound file plays.

Functions return a value, but you don't always have to use it. The sndSoundPlay function returns True if the sound plays and False if it doesn't. To keep the Elephant code simple, you aren't going to use the returned value.

To learn more about the sndPlaySound function and its options, go to msdn.microsoft.com and search for sndPlaySound.

Your elephant class is now complete. Granted, it's a pretty simple class, but it relies on the same coding principles you would use to create more robust classes. You are now ready to make elephant objects.

10 If you want to save your work, click the File menu in ArcMap and click Save As. Navigate to **C:\ArcObjects\Chapter09**. Rename the file **my_ex09a.mxd** and click Save. If you are continuing with the next exercise, leave ArcMap open. Otherwise close it.

Creating objects

Up to now, you have created objects with the aid of a user interface. For example, in chapter 3, you created a new form object from the Insert menu, and you then created CommandButton objects by dragging them from the Visual Basic Editor Toolbox onto the form.

Now you will create objects with code, by declaring and setting variables. Variables represent not only basic data types like numbers, dates, and strings, but can also represent objects.

Variables that represent basic data types are called intrinsic variables. You declare and set them with code like the following:

```
Dim X As Integer
X = 365
```

Then you use them.

```
MsgBox X
```

Variables that represent objects are called object variables. These variables are also declared and set, but the code is a little different. When you declare an object variable, you use the class name as the data type. For an elephant object, you would declare the variable as Elephant.

```
Dim E As Elephant
```

The line of code to set an object variable begins with the Set keyword. If you are creating a new object, you also use the New keyword between the equals sign and the class name. So to create a new elephant object, you would write the following line of code:

```
Set E = New Elephant
```

Now that you have an object variable called E, which refers to a new elephant, you can use the variable to set the elephant's properties and run its methods.

```
E.Name = "Mark"
E.Trumpet
```

Then, if you want, you can create some more elephants.

```
Dim E1 As Elephant
Dim E2 As Elephant

Set E1 = New Elephant
Set E2 = New Elephant
```

After creating them, you can set their properties so that each one is unique.

```
E1.Name = "Jerry"
E1.Age = 24

E2.Name = "Ron"
E2.Age = 28
```

What about objects you create through the user interface? You don't have to declare and set object variables for them because VBA does it for you. When you use Visual Basic Editor to create a form object, for example, and you set its Name property in the Properties window, that name becomes the form's variable name. Suppose you name the form frmXYZ. You can now proceed to set its properties and run its methods with code like the following:

```
frmXYZ.Show
```

Exercise 9b
In the previous exercise, you made an Elephant class and programmed its properties and methods. In this exercise, you will make new objects out of this class and use their properties and methods.

1 Start ArcMap and open **ex09b.mxd** in the **C:\ArcObjects\Chapter09** folder.

On the Standard toolbar you see the AddElephant button next to the Add Data button.

2 Right-click the AddElephant button and click View Source.

You see the AddElephant button's empty click event. You will write code there to create elephant objects from the Elephant class.

3 In the ThisDocument code module, in the AddElephant click event, add the following line of code to declare an object variable.

```
Dim theElephant As Elephant
```

4 Add the following line of code to create a new elephant object.

```
Set theElephant = New Elephant
```

After the Dim and Set lines run, a new object is in your computer's memory and you have a variable (theElephant) to refer to it.

5 Type **theelephant.** (including the dot).

Your Elephant class is recognized by VBA as a full-fledged class. As with any other object variable, after you type the dot you see the drop-down list of its properties and methods.

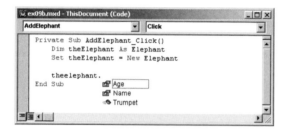

Next, you will write code to set this new elephant's properties.

6 Finish the line of code you started above by using an input box to set the elephant's name property.

```
theElephant.Name = InputBox("Enter name:")
```

7 Then set the elephant's age property.

```
theElephant.Age = InputBox("Enter age:")
```

When these two lines run, input boxes appear to your biologist-users. Next, you will write a line of code that displays the values back to the user.

8 Add a line of code to display the new elephant's name and age.

In the message box, the elephant's name will be on the top line. vbCrLf (carriage return line feed) will put the elephant's age on the second line.

```
MsgBox "Name: " & theElephant.Name & vbCrLf & _
       "Age: " & theElephant.Age, _
       vbInformation, "Add Elephant"
```

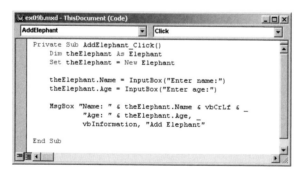

CHAPTER 9 • MAKING YOUR OWN OBJECTS

9 At the end of the AddElephant procedure, just before the End Sub line, add the following line of code to run the Trumpet method.

```
theElephant.Trumpet
```

Now you will test the AddElephant button.

10 Close Visual Basic Editor.

11 On the ArcMap Standard toolbar, click the AddElephant button.

12 In the first input box, type **Jack** for the name. Click OK.

13 In the second input box, type **35** for the age. Click OK.

You see the new elephant's data display.

14 Click OK on the Add Elephant message box.

After clicking OK, you hear the elephant trumpet. If you don't hear the trumpet, check your computer's volume, speakers, and sound card.

15 If you want to save your work, click the File menu in ArcMap and click Save As. Navigate to **C:\ArcObjects\Chapter09**. Rename the file **my_ex09b.mxd** and click Save. If you are continuing with the next chapter, leave ArcMap open. Otherwise close it.

Programming with interfaces

Using IApplication and IDocument
Using multiple interfaces

In the last chapter, you saw that objects come from classes and you learned how to make a VBA class of your own. In this chapter, you'll make the transition from working with VBA objects, like forms, combo boxes, and your own custom objects (the elephants), to working with ArcObjects, like maps and layers. To do that, you have to learn a new concept—the concept of a programming interface. This is something very different from a user interface.

Suppose you added a couple more properties and methods to the Elephant class from the last chapter and then decided to group them in some logical way. One group consists of properties and methods that are specific to elephants and the other group applies to animals in general.

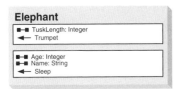

An *interface* is a logical grouping of properties and methods for a class. It might be based on the level of generality or on some other similarity of use or purpose.

The Elephant class in this example has two interfaces. By convention, interface names start with the letter "I," so these interfaces might be called IElephant and IAnimal. On a UML diagram, interfaces are shown with lollipops.

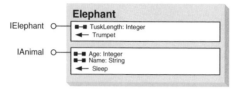

Elephant

IElephant ○── ■─■ TuskLength: Integer
                ◄── Trumpet

IAnimal ○── ■─■ Age: Integer
            ■─■ Name: String
            ◄── Sleep

You may be thinking that if that's all there is to an interface, it's pretty simple. In fact, there's more to it, but for this book, you don't need all the details. Briefly, interfaces are part of the Component Object Model (COM), a set of programming standards developed by Microsoft. A major benefit of COM is that it allows code written in one language, like Visual Basic, to work with code written in another language, like C++. The ArcObjects interfaces and classes, for example, are created with C++ and you can program with them using VBA.

In this book, you'll work with existing ArcObjects classes and interfaces, but you won't make any of your own. When you're ready to do that, you'll need a deeper understanding of interfaces. You can get that from the book *Exploring ArcObjects*.

Programming with classes that have interfaces is just a bit different from programming with classes that don't. Here's how you made an elephant object and set its Name property with the interfaceless Elephant class from the last chapter:

```
Dim e As Elephant
Set e = New Elephant
e.Name = "Ethan"
```

You didn't have to know anything about interfaces to use the elephant's Name property. As you work with classes that have interfaces, you have to find out which interface a property or method is on before you use it.

The Map class below has two interfaces: IMap and IActiveView. As you declare a Map variable, you have to choose one of its interfaces.

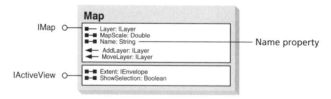

Map

IMap ○── ■── Layer: ILayer
            ■─■ MapScale: Double
            ■─■ Name: String ──────────── Name property
            ◄── AddLayer: ILayer
            ◄── MoveLayer: ILayer

IActiveView ○── ■─■ Extent: IEnvelope
                ■─■ ShowSelection: Boolean

Say you want to create a map and set its Name property, just as you did with the elephant a moment ago. The Name property is on the IMap interface. When you declare the Map variable, you use the interface name (IMap) instead of the class name (Map). So you declare the variable as IMap. The rest of the code is the same as the elephant code. The variable is set with the New keyword, and the Name property is set equal to a string.

```
Dim pMap As IMap
Set pMap = New Map
pMap.Name = "Ryan's Map"
```

Had you wanted the Map class's Extent property, you would have declared the variable as IActiveView, since Extent is on the IActiveView interface.

```
Dim pActiveView As IActiveView
Set pActiveView = New Map
pActiveView.Extent = someNewExtent
```

You only have access to the properties and methods of the interface to which a variable is declared. So with pMap you only have access to IMap's properties and methods, and with pActiveView you only have access to IActiveView's properties and methods. For now you will use one interface at a time. In the second exercise of this chapter, you'll learn how to use more than one interface at a time.

Variables like pMap and pActiveView are called *pointer* variables because they point to a specific interface. You can say that pMap points to the IMap interface and pActiveView points to the IActiveView interface. By convention, pointer variable names have a "p" prefix.

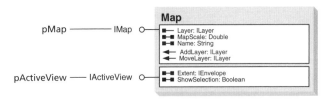

You may be wondering why you haven't had to use interfaces with VBA objects. Don't they have them? As a matter of fact they do, but whereas ArcObjects interfaces are exposed, VBA interfaces are hidden by the programmers who created them. When you use VBA objects, variables point to interfaces, but it's done for you behind the scenes.

That's an advantage in one way, because you don't have to think about it, but it's a disadvantage in another way. When interfaces are exposed, they can be shared. That means you can create your own custom classes and have those classes use interfaces created by someone else.

For example, you can create a class and tell it to use the ArcObjects IContentsView interface. The tabs in the ArcMap table of contents each have the IContentsView interface. When your class uses that ArcObjects interface, it fits in with other ArcObjects. IContentsView on your object is recognized by ArcMap and becomes another tab in the ArcMap table of contents. The custom tab shown below also has some ArcCatalog interfaces. The result is a new tab that contains a fully functional ArcCatalog tree.

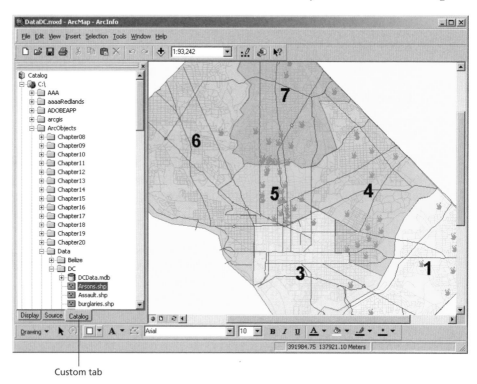

Custom tab

When interfaces are hidden, you can't use them to make your own custom classes. In this book, you won't reap the benefits of exposed interfaces—that harvest will have to wait until you start programming your own ArcObjects.

Using IApplication and IDocument

Normally, when you use objects, you declare and set variables for them. When you start ArcMap and open a map document, however, a couple of objects are already in use. This is because you can't do anything at all until you have a document open in ArcMap.

The two special objects already in use are the Application object and the MxDocument object. Since these objects are in use, it follows that variables have already been declared for them. The name of the Application object variable is Application. The name of the MxDocument object variable is ThisDocument. These two are the only predefined object variables there are.

Application

ThisDocument ——

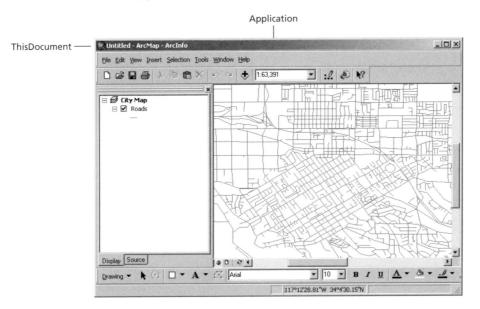

All ArcObjects classes have interfaces; therefore, all ArcObjects variables are pointer variables that point to interfaces. Application and ThisDocument point to interfaces on the Application and MxDocument classes. (These two special variables don't observe the convention of beginning with a lowercase "p.")

The Application variable points to the IApplication interface on the Application class. The variable has access to all the IApplication properties and methods, among which is the Caption property.

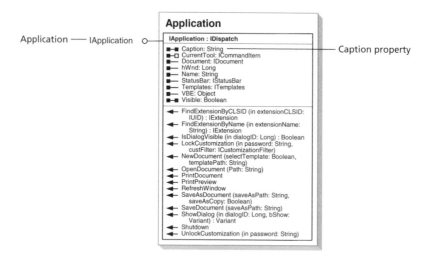

Application — IApplication

Caption property

The code below uses the predefined Application variable to set the ArcMap window's caption.

```
Application.Caption = "Save Elephants"
```

In class diagrams, the barbells next to a property tell you whether you can get or set the property. A two-sided barbell, like the one by the Caption property, means that you can both get and set the property. A left-sided barbell, like the one by the Name property, means you can get the property, but you can't set it to something new. A right-sided barbell means that you can set the property but not get it.

The ThisDocument variable points to the IDocument interface on the MxDocument class. The variable has access to all the IDocument properties and methods, among which is the Title property.

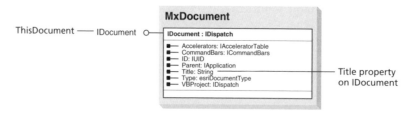

ThisDocument — IDocument

Title property on IDocument

In the diagram, all of IDocument's properties have left-sided barbells, which means you can only get their values. Say you want to display the title of the current map document in a message box. The Title property is shown as *Title: String,* which means that the Title property returns a string. You can display the map document's file name with the following code. No need to declare or set the ThisDocument variable.

```
Msgbox ThisDocument.Title
```

Exercise 10a

You have been asked to make some changes to the Washington, D.C., Police Department's Crime Analysis toolbar (the one you worked on in chapter 6). In this exercise, you will add some security to the map document—since it contains sensitive data—and you'll give the application a more meaningful title.

Because the crime map contains data that should only be seen by certain detectives, you'll write an If Then statement to check the user's identity. As the crime map opens, your code will ask for a password. Users who know it are authorized to open the map; others will get a warning and be sent to a less sensitive city street map.

When ArcMap starts up, its title bar contains three elements: the name of the .mxd file, the name ArcMap, and the name of the current software license (ArcView, ArcEditor, or ArcInfo). Your first task will be to use the Application object's Caption property to set a title that will be more meaningful to police personnel.

Application title

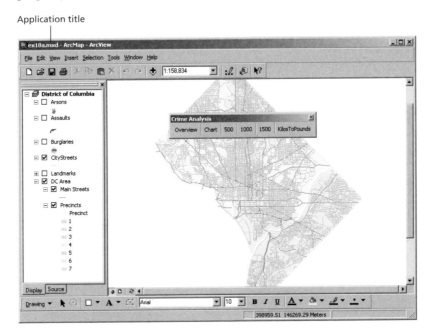

1 Start ArcMap and open **ex10a.mxd** in the **C:\ArcObjects\Chapter10** folder.

You see the District crime map, and you see that the ArcMap title is "ex10a.mxd - ArcMap - ArcView" (or ArcEdit or ArcInfo depending on your license).

2 Click the Tools menu, point to Macros, and click Visual Basic Editor.

3 Unless the ThisDocument code module is already open, go to the Project window. Under Project (ex10a.mxd), under ArcMap Objects, double-click ThisDocument.

The ThisDocument code window opens, and you see the code you wrote in chapter 6 for the buttons on the Crime Analysis toolbar.

4 Click the object list drop-down arrow and click MxDocument. Click the procedure list drop-down arrow and click OpenDocument.

The OpenDocument event is to the map document as the Initialize event is to the UserForm. These two events happen just before the user sees a map document or a dialog box.

The code you are about to write in the OpenDocument event will ask the user for a password. If they know the password, the map will open and the ArcMap title will be replaced by a new one. Otherwise, a warning will appear, and a less sensitive street map will open.

5 In the OpenDocument event, add two lines of code to declare a string variable and set it with an input box that asks the user for the password.

```
Dim strPassword As String
strPassword = InputBox("Enter password")
```

You will set up an If Then statement to verify the password.

6 In the OpenDocument event, enter the following If Then statement.

```
If "Carter" = strPassword Then

Else

End If
```

Next, you will write code to run when the statement is true. You will use the predefined Application variable and its Caption property to change the ArcMap application title.

7 After the Then and before the Else, type **Application.** (including the dot).

After typing the dot, you see a list of Application's properties and methods.

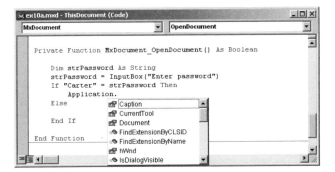

Since the Application variable points to the IApplication interface, the properties and methods in the drop-down list above match those in the IApplication diagram.

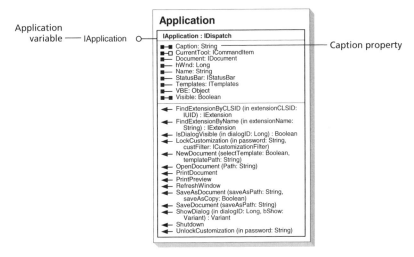

8 Finish the line above by setting the Caption property to District Crime.

```
Application.Caption = "District Crime"
```

When the wrong password is entered, your code will warn the unauthorized user that they can't open the crime map's .mxd file. To make the warning message, you will use the IDocument interface's Title property to get the .mxd file name.

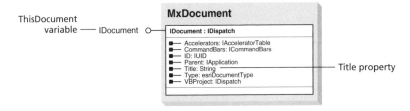

9 After the Else keyword, add the following code to concatenate the map document's title with the warning instructions.

```
Msgbox ThisDocument.Title & " is a secure map. " & _
    "A street map will be opened for you."
```

To send unauthorized users into another map, you will use the IApplication's NewDocument method. Given a path to a map template (.mxt file), this method opens the specified map. Because NewDocument is on the IApplication interface, you can still use the predefined Application variable.

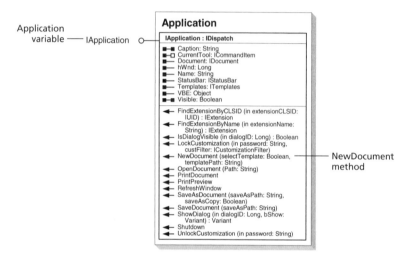

10 Add the following line of code that uses the NewDocument method and a path to the map template file. (If you installed the data at another location, type the appropriate path.)

The streets.mxt file in the Data folder contains the District's boundaries and streets but no sensitive crime data.

```
Application.NewDocument _
    False, _
    "C:\ArcObjects\Data\dc\streets.mxt"
```

NewDocument's first argument is either true or false. If true, the New template dialog box opens before a map opens. Your users don't need to see any dialog boxes, so you use false.

The second argument is the path to the template .mxt file. Whenever you type path and file names in code, you should double-check your typing and use a file browser or ArcCatalog to confirm the file's location. Second only to typos, incorrect path and file names are probably the greatest source of VBA errors.

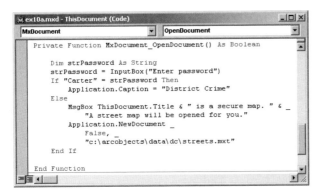

```
ex10a.mxd - ThisDocument (Code)                              _|□|×|
MxDocument                    ▼    OpenDocument              ▼

    Private Function MxDocument_OpenDocument() As Boolean

        Dim strPassword As String
        strPassword = InputBox("Enter password")
        If "Carter" = strPassword Then
            Application.Caption = "District Crime"
        Else
            MsgBox ThisDocument.Title & " is a secure map. " & _
                "A street map will be opened for you."
            Application.NewDocument _
                False, _
                "c:\arcobjects\data\dc\streets.mxt"
        End If

    End Function
```

The OpenDocument event procedure's code runs when a map document opens. So to test it, you will save your work, close the map document, and then reopen it.

11 Close Visual Basic Editor.

12 In ArcMap, click the File menu and click Save As. Navigate to **C:\ArcObjects\Chapter10**. Rename the file **my_ex10a.mxd** and click Save.

Instead of quitting and restarting ArcMap, you will open a blank map document and then reopen the crime map you just saved.

13 In ArcMap, click the New Map File button on the Standard toolbar.

A new map document opens. To simulate an unauthorized user, you will try to open the crime map with an incorrect password.

14 Click the File menu and click Open. Open **my_ex10a.mxd** in the **C:\ArcObjects\Chapter10** folder.

15 For the password, type **ArcObjects**. Click OK.

A message warns that a streets map will be opened.

```
ArcMap                              ×

my_ex10a.mxd is a secure map. A street map will be opened for you.

              [    OK    ]
```

16 Click OK on the warning message.

A new map opens with the streets and precincts layer. No sensitive data layers are shown.

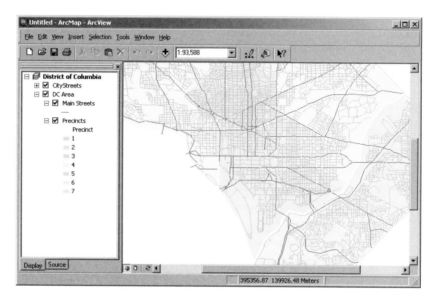

Next, you will open the crime map with the correct password.

17 Click the File menu and click Open. Open **my_ex10a.mxd** in the **C:\ArcObjects\Chapter10** folder. If prompted to save changes, click No.

18 For the password, type **Carter**. Your If Then statement is case sensitive, so make sure the "C" in Carter is uppercase. Click OK.

The crime map document opens. The application title is now District Crime.

19 If you are continuing with the next exercise, leave ArcMap open. Otherwise close it.

If you leave ArcMap open, the application title will remain District Crime. The title stays until you run code to change it, or until you quit ArcMap.

Using multiple interfaces

In the previous exercise, you used the Application and ThisDocument predefined variables. The interfaces they point to, IApplication and IDocument, don't have many properties or methods. It's great to have these two variables as starting points, but there are other interfaces out there with lots of other properties and methods that you might find useful.

As you know, ArcObjects classes can have several interfaces. If you have a variable pointing to one interface on an object, you can write two lines of code to declare and set a second variable to point to one of that object's other interfaces. You end up with two variables and access to the properties and methods of two interfaces.

In the example below, you are working with the Dog class. You want to create a new dog and get it to bark. The Bark method is on the IDog interface, so you declare a variable to that interface. In accordance with convention, you name the variable pDog1. Then you use Set and New to create a new dog object. Finally, you run the Bark method with a 2 to make the dog bark twice.

```
Dim pDog1 As IDog
Set pDog1 = New Dog
pDog1.Bark 2
```

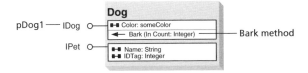

Say you want to create a second dog and name it Rex. The Name property is on the IPet interface, so you declare a variable to that interface and you call it pPet2. You use Set and New to create the second dog and then, with a third line of code, you set its Name property to Rex.

```
Dim pPet2 As IPet
Set pPet2 = New Dog
pPet2.Name = "Rex"
```

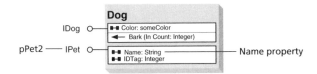

You now have two dogs. The first dog, referred to with the pDog1 variable, is nameless but barked twice. The second dog, referred to with the pPet2 variable, is named Rex and has not yet barked.

pDog1 ——

pPet2 ——

Rex

Now what if you want Rex to bark? The pPet2 variable points to the IPet interface, which only gets you to the Name and IDTag properties.

You might be wondering if you can use the pDog1 variable, and write:

```
pDog1.Bark
```

You can, but that will only make the nameless dog bark. What you need to do is set up a second variable that points to Rex's IDog interface. This switching of interfaces takes two lines of code: one to declare a variable to the second interface and another to set that variable equal to the variable you already have. Switching interfaces is called QueryInterface, or QI for short.

To get Rex to bark, you declare a second pointer variable to the IDog interface, which has the Bark method.

```
Dim pDog2 As IDog
```

Instead of setting the variable equal to a New Dog (which would create a third dog), you set it equal to pPet2.

```
Set pDog2 = pPet2
```

Now the pPet2 and pDog2 variables both refer to the same object (Rex); each, however, points to a different interface.

Since pDog2 points to the IDog interface, you can use the Bark method. The next line of code gets Rex to bark three times.

```
pDog2.Bark 3
```

The pPet2 variable points to Rex's IPet interface. So you could use pPet2 to set Rex's IDTag property with the line of code below.

```
pPet2.IDTag = 714
```

You have created a dog, named him Rex, got him to bark, and given him an IDTag number. Now how about getting back to the other dog and giving it a name? The variable pDog1 points to the nameless dog's IDog interface, but the Name property is on the IPet interface. So to set the dog's name, you have to switch interfaces. This time you have to declare a new variable that points to the IPet interface, then set that variable equal to the one you already have.

```
Dim pPet1 As IPet
Set pPet1 = pDog1
```

Both pDog1 and pPet1 refer to the nameless dog, but point to different interfaces. You can now set the nameless dog's Name property with the pPet1 variable.

```
pPet1.Name = "Radar"
```

And, of course, you can still make Radar bark (again) with the original pDog1 variable.

```
pDog1.Bark 4
```

In all, you have four variables referring to the two dogs. You can use any of the four variables with the properties and methods on their respective interfaces.

pDog1 ——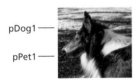

pPet1 ——

pDog2 ——

pPet2 ——

Exercise 10b

Over time, the crime map will be updated and used by many different people. In this exercise, you will write code to keep track of which users open it. This will help you see which members of the department are getting the most value from the map.

You'll store the information in the map document's Properties dialog box, which has input boxes for things like the map's title, subject, and author. Specifically, your code will write each user's password and access time into the map's Comments area.

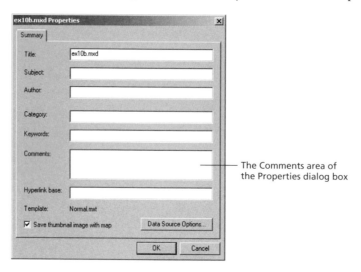

The Comments area of the Properties dialog box

Also, for the benefit of people sitting in the back of the room, you'd like to make the crime map as big as you can. At the start of this exercise, you'll write code to make the ArcMap window fill the display screen.

1 Start ArcMap and open **ex10b.mxd** in the **C:\ArcObjects\Chapter10** folder.

2 For the password, type **Carter**. Click OK.

You see the District Crime map.

3 Click the Tools menu, point to Macros, and click Visual Basic Editor.

The current password code is located in the OpenDocument event procedure in the ThisDocument code module. You'll go there to write more code.

4 Unless ThisDocument is already open, go to the Project window under Project (ex10b.mxd) and double-click ThisDocument under ArcMap Objects.

The ThisDocument code window opens and you see your code from earlier exercises.

5 Click the object list drop-down arrow and click MxDocument. Click the procedure list drop-down arrow and click OpenDocument.

You see the If Then statement that checks for the user's password.

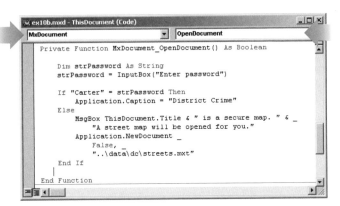

```
ex10b.mxd - ThisDocument (Code)
MxDocument                        OpenDocument

    Private Function MxDocument_OpenDocument() As Boolean

        Dim strPassword As String
        strPassword = InputBox("Enter password")

        If "Carter" = strPassword Then
            Application.Caption = "District Crime"
        Else
            MsgBox ThisDocument.Title & " is a secure map. " & _
                "A street map will be opened for you."
            Application.NewDocument _
                False, _
                "..\data\dc\streets.mxt"
        End If

    End Function
```

In the last exercise, you used the predefined Application variable, which points to the Application class's IApplication interface. In this exercise, you will switch to the IWindowPosition interface, also on the Application class. IWindowPosition has properties for controlling the size and position of the ArcMap window.

When class diagrams are shown in this book, some of their interfaces, properties, and methods may be omitted to save space. For example, the Application class has ten interfaces, but only two are shown in the following diagram. Also, all of the IApplication interface's methods have been omitted.

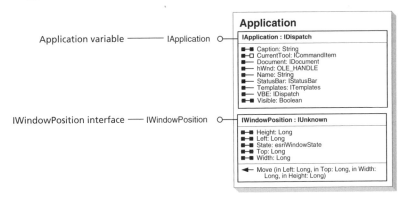

Application variable ——— IApplication

Application

IApplication : IDispatch
- Caption: String
- CurrentTool: ICommandItem
- Document: IDocument
- hWnd: OLE_HANDLE
- Name: String
- StatusBar: IStatusBar
- Templates: ITemplates
- VBE: IDispatch
- Visible: Boolean

IWindowPosition interface ——— IWindowPosition

IWindowPosition : IUnknown
- Height: Long
- Left: Long
- State: esriWindowState
- Top: Long
- Width: Long

- Move (in Left: Long, in Top: Long, in Width: Long, in Height: Long)

To get to the properties for setting the window size, you need a variable that points to the IWindowPosition interface. You can start from the predefined Application variable, but then you'll need to use QueryInterface to switch from IApplication to IWindowPosition. This means declaring a variable for IWindowPosition and setting it equal to the Application variable.

6 After the If Then line, but before the Application.Caption line, add the following line of code to declare a variable that points to the IWindowPosition interface.

```
Dim pWindow As IWindowPosition
```

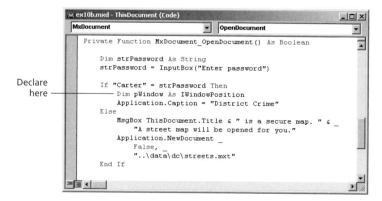

Declare here ——

```
ex10b.mxd - ThisDocument (Code)

MxDocument                          OpenDocument

    Private Function MxDocument_OpenDocument () As Boolean

        Dim strPassword As String
        strPassword = InputBox("Enter password")

        If "Carter" = strPassword Then
            Dim pWindow As IWindowPosition
            Application.Caption = "District Crime"
        Else
            MsgBox ThisDocument.Title & " is a secure map. " & _
                "A street map will be opened for you."
            Application.NewDocument _
                False, _
                "..\data\dc\streets.mxt"
        End If
```

7 Below the Dim statement, add the following line of code to set the variable equal to Application.

```
Set pWindow = Application
```

You now have two variables pointing to two different Application interfaces, as shown in the following graphic:

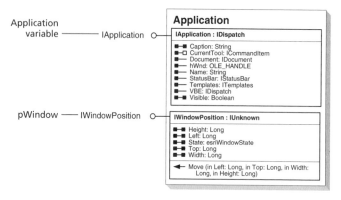

Application variable ——— IApplication ○—

pWindow ——— IWindowPosition ○—

Application

IApplication : IDispatch

■—■ Caption: String
■—□ CurrentTool: ICommandItem
■—■ Document: IDocument
■—■ hWnd: OLE_HANDLE
■—■ Name: String
■—■ StatusBar: IStatusBar
■—■ Templates: ITemplates
■—■ VBE: IDispatch
■—■ Visible: Boolean

IWindowPosition : IUnknown

■—■ Height: Long
■—■ Left: Long
■—■ State: esriWindowState
■—■ Top: Long
■—■ Width: Long

◄—— Move (in Left: Long, in Top: Long, in Width: Long, in Height: Long)

8 Add the following line of code to set the Application window's State property to esriWSMaximize.

```
pWindow.State = esriWSMaximize
```

This code will maximize the ArcMap window for the District Crime application. To learn more about the options for the State property, you could highlight State in the code window and press F1 to open the ArcObjects developer help window.

Next you will write code to track the map's users. As you recall, you'll do this by writing the user's password and access time into the map document's Properties dialog box. All the input boxes on this dialog box are controlled by properties on the MxDocument class's IDocumentInfo interface, shown below. You will set the Comments property.

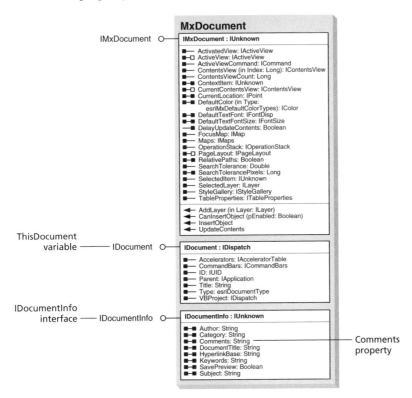

Once again, you will start with a predefined variable and use QueryInterface. This time, the predefined variable is ThisDocument and you are switching from the IDocument interface to the IDocumentInfo interface. To make the switch, you'll declare a variable for IDocumentInfo and set it equal to the ThisDocument variable.

9 In the OpenDocument event, after the strPassword variable is set with the input box and before the If Then statement, add the following line of code to declare a variable as IDocumentInfo.

```
Dim pInfo As IDocumentInfo
```

Declare here ———

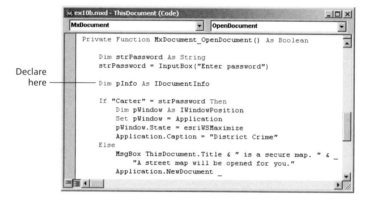

```
ex10b.mxd - ThisDocument (Code)

MxDocument                    OpenDocument

Private Function MxDocument_OpenDocument() As Boolean

    Dim strPassword As String
    strPassword = InputBox("Enter password")

    Dim pInfo As IDocumentInfo

    If "Carter" = strPassword Then
        Dim pWindow As IWindowPosition
        Set pWindow = Application
        pWindow.State = esriWSMaximize
        Application.Caption = "District Crime"
    Else
        MsgBox ThisDocument.Title & " is a secure map. " & _
            "A street map will be opened for you."
        Application.NewDocument _
```

10 After the Dim statement, add a line of code to set the variable equal to ThisDocument.

```
Set pInfo = ThisDocument
```

You now have two variables pointing to two different interfaces. As shown in the simplified MxDocument class diagram below, ThisDocument points to the IDocument interface and pInfo points to the IDocumentInfo interface.

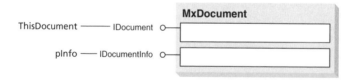

MxDocument

ThisDocument ——— IDocument

pInfo ——— IDocumentInfo

11 In the If Then statement, after the pWindow.State line but before the Application.Caption line, add the following line of code. It writes the current time, the user's password, and the word Authorized into the map document's comments area.

```
pInfo.Comments = pInfo.Comments _
                & " " & Now _
                & " " & strPassword _
                & " Authorized" & vbCrLf
```

The empty quotes separate the entries and put spaces between the time and the user name. You won't see what the full comment looks like until after the code runs and you have a chance to open the Properties dialog box.

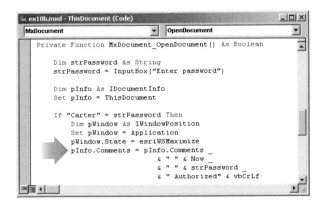

To guarantee that the comments are saved into the .mxd file, you will add a line of code to save the file automatically. Otherwise, a user could open the map document and quit without saving, and their name wouldn't be added to the comments. You will use IApplication's SaveDocument method, shown below.

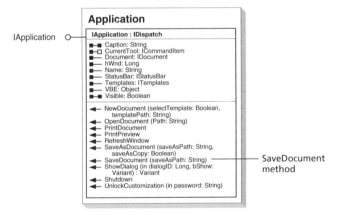

12 Add the following line of code to save the map document.

```
Application.SaveDocument
```

You would also like comments to be added when an unauthorized user tries to open the map. You'll add that code next.

13 After the Else keyword, but before the ThisDocument warning message, add the following line of code. It writes the current time, the user's password, and the word Unauthorized into the map document's comments area.

```
pInfo.Comments = pInfo.Comments _
                 & " " & Now _
                 & " " & strPassword _
                 & " Unauthorized" & vbCrLf
```

14 Add the following line of code to save the unauthorized user's information.

```
Application.SaveDocument
```

When the user enters an incorrect password, their password and time are added to the comments area and the SaveDocument method runs to save the .mxd file. Then the warning appears and the NewDocument method runs to open the streets.mxt file.

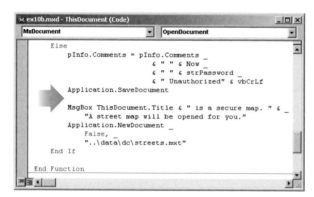

Because your code is in the map document's OpenDocument event, you need to open the document to test it. So you'll save your work, close the map document, open a new one, and then reopen this one.

15 Close Visual Basic Editor.

16 In ArcMap, click the File menu and click Save As. Navigate to **C:\ArcObjects\Chapter10**. Rename the file **my_ex10b.mxd** and click Save.

17 In ArcMap, click the New Map File button on the Standard toolbar.

A new empty map document opens. Next, you will try to open the crime map, but with an incorrect password to simulate an unauthorized user.

18 Click the File menu and click Open. Open **my_ex10b.mxd** in the **C:\ArcObjects\Chapter10** folder.

19 For the password, type **ArcObjects**. Click OK.

20 Click OK on the warning message.

A new map opens with the streets and precincts from the Streets.mxt template. Next, you will open the document with the correct password.

21 Click the File menu and click Open. Open **my_ex10b.mxd** in the **C:\ArcObjects\Chapter10** folder. If prompted to save changes, click No.

22 For the password, type **Carter**. Click OK.

The District Crime document opens and its window is maximized.

23 Click the File menu and click Map Properties (or Document Properties).

In the Comments area, you see that the first user was the unauthorized ArcObjects and the second user was the authorized Carter.

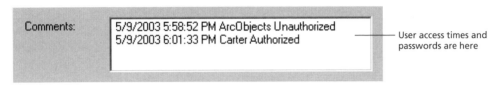

User access times and passwords are here

24 Click Cancel to close the Properties dialog box.

Now that you have programmed with interfaces you are ready to learn about another COM rule: Once an interface is published it won't be altered in future releases of the software. That means an interface will always have the same properties and methods. Properties and methods will always have the same arguments and return values. Arguments and return values will always be of the same data type.

If ESRI programmers want to enhance something about the IMap interface in a future release of the software, they never change IMap. Instead they create another interface called IMap2 and make changes to it.

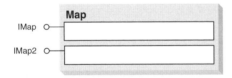

With this development approach, your existing code that works with the Map class and the IMap interface will continue to work in the future with the next version of the software. At that time, if you like, you could edit your code to point to the new interface, but it's all up to you.

25 If you are continuing with the next chapter, leave ArcMap open. Otherwise close it.

ABSOLUTE AND RELATIVE PATHS

As you wrote code for this exercise, you may have noticed that the path to the streets.mxt file was different than the one you typed in the previous exercise.

In exercise 10a, you typed:

```
"C:\ArcObjects\Data\dc\streets.mxt"
```

This is a hard-coded (or absolute) path. The path begins with a drive letter and lists each subsequent folder that leads to the streets.mxt file. A hard-coded path will locate the target data as long as it is stored in the specified location. If the target data is stored somewhere else (if it is moved, for example), the path will break and the data will not be found.

In exercise 10b, the hard-coded path has been changed to:

```
"..\Data\dc\streets.mxt"
```

This is a relative path. It leads from the file that contains the code (in this case, ex10b.mxd) to the target data. The two dots mean "go up one folder in the directory structure." Since ex10b.mxd is in the Chapter10 folder, the dots direct the code up one level to the ArcObjects folder. From there, the rest of the path is hard-coded.

You can go up as many levels as you want by using sets of dots:

```
"..\..\..\streets.mxt"
```

Relative paths make your code and data more portable. You can move the ArcObjects folder to any location you want. The code will always find streets.mxt, because the target data stays in the same location relative to ex10b.mxd—up one folder to ArcObjects, then down to Data, and down to dc.

When you know that your data structure isn't going to change, use hard-coded paths. For data that may move, use relative paths.

Navigating object model diagrams

Getting layers

Creating and assigning colors

So far, as you've worked with objects—getting and setting their properties, running methods, even switching interfaces—you've worked with one class at a time. But you know, of course, that there are hundreds and hundreds of ArcObjects classes.

The ArcObjects you're familiar with, like Application and ThisDocument, are always right there waiting for you, like a limousine at your door. Typically, however, you need to work with objects that are not quite so obligingly at your service. For example, suppose you want to program a button to change the color of a Rivers layer. A layer's color is an object (created from a class like RGBColor below) with properties to mix different amounts of red, green, and blue.

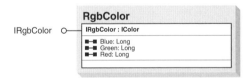

Even though you have never used RGBColor before, you know from the last two exercises that to make any new object you declare a variable to one of its interfaces, you set the variable to a new object with the New keyword, and after that you can get and set properties and run methods.

```
Dim pColor As IRGBColor
Set pColor = New RGBColor
pColor.Blue = 180
```

But your goal is to make the Rivers layer blue. How does the code know which layer, in which data frame, you want to change? It's a different situation when, instead of creating new objects, you want to do something with an object that's already in play somewhere in your application. To do that, you have to find your way to the object.

It's kind of like hopping to a marker in a game of hopscotch, or swinging from one vine to another in the jungle. The hopscotch squares (or the vines) are classes, and you hop from one to the next until you get to the one you want.

For example, to change the River layer's color, you would make your way from the MxDocument class (MxDocument is the map document or currently opened .mxd file), to the Map class (the Map class in ArcObjects refers to a data frame), to the FeatureLayer class, to the Renderer class (renderer is a fancy word for a layer's legend), to the Symbol class, and to the Color class, and there, at last, you would change the properties of the color object.

MxDocument → Map → Layer → Renderer → Symbol → Color

Now, because you have to make a particular series of hops to get from your starting point to your destination, you can infer that there are certain special connections or relationships between classes. The MxDocument class is *associated* with the Map class, or connected to it, and that's why you can go from one to the other. However, the MxDocument class is not connected to the Color class. And therefore you can't go straight from MxDocument to Color.

In UML, associated classes are connected with a solid gray line. The line means you can get from one class to another. This relationship can usually be expressed in plain English by saying that objects in one class have objects in the connected class. For example, the MxDocument class is connected to the Map class, and it's also natural to say that map documents have data frames. The Symbol class is connected to the Color class, and it's also natural to say that symbols have colors.

Other symbols on the diagram give you additional information about how the classes are related. The asterisk by the Map class means multiplicity. In other words, a map document can have many maps in it. The relationship between the Application class and the MxDocument class would not have an asterisk because you can't show two .mxd files in one ArcMap window.

The black diamond tells you that the object with the diamond is composed of the associated objects. You can think of the object with the diamond as a container. The map document contains maps. If you delete the map document, any maps inside it are deleted too.

Navigating object model diagrams means hopping around from one class to another. Now we'll consider how this hopping is accomplished technically.

Classes have properties on their interfaces. When you set a property, you have to use the right type of value. For example, on the ILayer interface of the Layer class below, there is a Name property and its value must be set to a string. ILayer also has a Visible property, which must be set to a Boolean value (true or false).

When you get property values you have to be ready to receive the right type of value. A programmer would say that the Name property *returns* a string and the Visible property *returns* a Boolean.

Some properties return interfaces. For example, the MxDocument class below has a property called FocusMap that returns an IMap. (FocusMap tells you which data frame in the document is active.) What this means is that FocusMap doesn't return a map object, but it returns a pointer to a map object's IMap interface. Through that pointer, you have access to the object's IMap properties and methods.

On the diagram, a line connecting two classes tells you that you can get one object (well, the interface of the object) from another. To make the connection, a property on the one object will return an interface of the connected object. Here the FocusMap property on IMxDocument returns the IMap interface of the connected map object. If you want to hop from the MxDocument class to the Map class, you get the FocusMap property.

Let's look at how to write the code that takes you from MxDocument to Map. Your starting point is the predefined ThisDocument variable, which points to the IDocument interface on MxDocument.

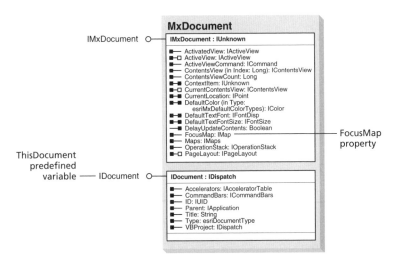

Since the FocusMap property is on IMxDocument, you first have to use QueryInterface to switch interfaces. Declare a variable to the interface you need, and set it equal to the variable you already have.

```
Dim pMxDoc As IMxDocument
Set pMxDoc = ThisDocument
```

Now that you have a variable that points to IMxDocument, you can get the FocusMap property. Since this property returns an IMap interface, you declare a new variable to IMap.

```
Dim pMap As IMap
Set pMap = pMxDoc.FocusMap
```

That's it. You've just hopped from MxDocument to Map. If you wanted to go on to work with a layer object, you would do the same thing: find a property in the Map class (IMap has a Layer property) that returns an interface of the Layer class (the Layer property returns ILayer). Since QueryInterface isn't needed here, it only takes two lines of code to go from Map to Layer.

```
Dim pLayer As ILayer
Set pLayer = pMap.Layer(1)
```

As long as classes are associated (connected with a line on an object model diagram), there will be some property that takes you from one to the other.

Maybe you are wondering how you're supposed to know which classes are associated and which are not? You find out by looking at object model diagrams, which are poster-sized drawings that show not only classes with their properties and methods, but also the relationships between classes.

A single diagram is not big enough to hold all the ArcObjects classes and their interfaces. In fact, there are many diagrams, organized by categories of classes. For example, the Geometry diagram contains classes for points, lines, and polygons. The Geodatabase diagram contains classes for tables, feature classes, rows, and fields. If a class on one diagram is connected to a class on a different diagram, you will be referred to that diagram (just as an atlas may refer you to a map on another page).

The following object model diagram shows more than a hundred classes, one of which is the Map class and its interfaces.

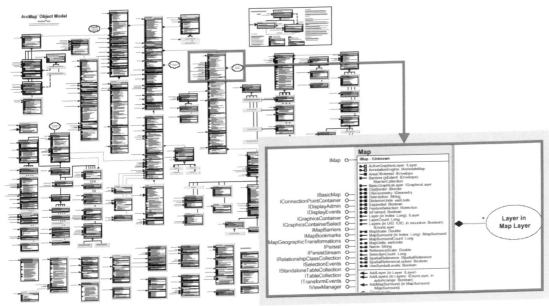

The oval indicates that the Map class is associated with the Layer class, which is found on the Map Layer diagram.

Object model diagrams are pretty detailed, but they only use about eight different symbols. Once you learn these symbols, and get a little practice, you'll be able to follow a diagram as easily as you would a street map. (Or maybe more easily!)

Object model diagrams are located in the developer help. If you don't have the help loaded on your computer, you can get the diagrams from the online developer help (arcobjectsonline.esri.com). If you have some wall space, you can buy printed posters from ESRI (www.esri.com/ExploringArcObjects). And, finally, all the diagrams are included on the *Getting to Know ArcObjects* data CD as PDF files.

You now have the central idea about how to navigate ArcObjects classes. As usual, it's not quite the whole story. The association relationship that you've been learning about is actually just one of three class relationships. Now we'll talk a little bit about the other two.

Instantiation, also called the Creates relationship, is a relationship in which one class has a method that creates new objects from another class. Imagine a Shoemaker class with a CreateShoes method. Running this method creates a new object from the Shoes class and returns you its IShoes interface. In UML, this relationship is shown by a dashed black line with an arrow that points to the created object. You won't instantiate any objects in this chapter, but you will later in the book.

The third type of class relationship is *inheritance*. Inheritance is when a particular class uses an interface (programmers say "implements" an interface) from a more general class.

Remember when in the last chapter we talked about grouping the Elephant class's properties onto different interfaces? Properties specific to elephants went on the IElephant interface, and properties that applied to all animals went on IAnimal, where they could be used by the Giraffe class, the Moose class, and so on. On an object model diagram, it wouldn't be convenient to draw the IAnimal interface on every class that used it—it would take up too much space. So programmers make a so-called abstract class, and call it something like Animal, and draw the IAnimal interface on this class.

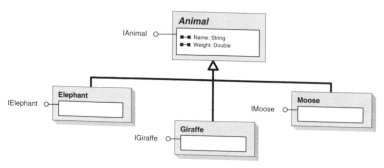

The point here is that Animal isn't a full-blooded class. There are no Animal objects: there are only Elephant objects, Giraffe objects, Moose objects, and so on. The Animal class is nothing more than a schematic convenience, a parking place for an interface that is used by many different classes.

Inheritance is shown with solid black lines and a triangle. Below, the triangle is on the Layer class, which means that the other connected classes inherit all of Layer's interfaces (which in this case is just ILayer).

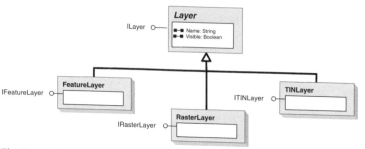

The FeatureLayer, RasterLayer, and TINLayer classes all inherit from the Layer class. This means that all three classes implement the ILayer interface, which is drawn on the abstract Layer class. The triangle points to the abstract class (also called the superclass).

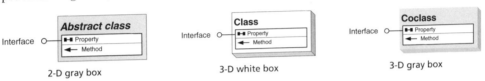

Since you're going to work with abstract classes in this chapter, let's see how you write code for them. Suppose you want to create a new FeatureLayer and set its Name property. The Name property is on the ILayer interface, which is shown on the abstract Layer class. But since the FeatureLayer class inherits this interface, you write your code as if you were seeing the interface on the FeatureLayer class.

```
Dim pLayer As ILayer
Set pLayer = New FeatureLayer
pLayer.Name = "USA"
```

This leads to a final point. The fact that there are different class relationships means that there are different ways to diagram classes. (You might have noticed this in the previous diagrams.)

Abstract classes, like the Layer class you were just looking at, are drawn as 2-D gray boxes. As mentioned before, no objects come out of abstract classes. They are just parking spots for common interfaces.

Classes (which it's sometimes convenient to call regular classes, to distinguish them from classes in general) are drawn as 3-D white boxes. Objects from regular classes have to be instantiated—you can't create them yourself with the New keyword. If you want an object from a regular class, you have to find another class that knows how (has a property or method) to make or get these objects.

Coclasses are drawn as 3-D gray boxes. These are the kinds of classes you're already familiar with. You can create objects from coclasses with the New keyword. You can also get these objects by using properties of other objects that return them.

Getting layers

In this exercise, you'll write some code that gets the layers in a data frame and turns them off. This involves navigating from a starting point through several classes to get to the layer objects you want. You'll follow the path shown below, except that you'll start with MxDocument (where the predefined ThisDocument variable is waiting for you) and skip the step of hopping from Application to MxDocument.

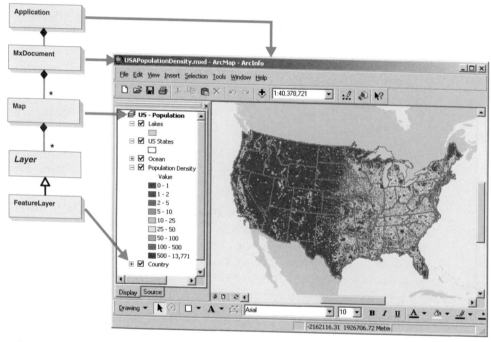

Along the way, you'll do a couple of fun things. You'll write a For Each loop that cycles through each layer in the data frame to turn it off. And you'll set object properties to update the table of contents and the map display after the layers have been turned off.

You may be wondering how you'll ever learn all the class relationships on your own and be able to navigate easily from one class to another. It can be a little overwhelming at first, but it's like learning your way around a network of hiking trails. You make some wrong turns, get lost a few times, but then you develop a sense of how everything is connected and it starts to become automatic.

Surprisingly, you probably know more than you think about class relationships, because you have been an ArcGIS user. For example, you might not need the diagram above to tell you that Maps contain Layers, because in your ArcGIS experience you have added hundreds of layers to maps. You already know they have a relationship. If you've ever deleted a map, you know the layers get deleted too.

Exercise 11a

During a crime analysis session, you turn many crime layers on. To get a fresh start on analyzing the next crime, you turn them all off again. To save a bit of time, you are going to create a button to do that for you. Before writing the code, you'll look at a couple of object model diagrams to see the class relationships. This means you'll need Adobe® Acrobat Reader®. If you don't have it, you can download the latest version for free at www.adobe.com.

1 Start Acrobat Reader and open **ArcMap Object Model.pdf** in the **C:\ArcObjects \Diagrams** folder.

Maps sometimes have a big red dot that says "You are here." On the ArcMap diagram, MxDocument (with its ThisDocument variable) is your big red dot.

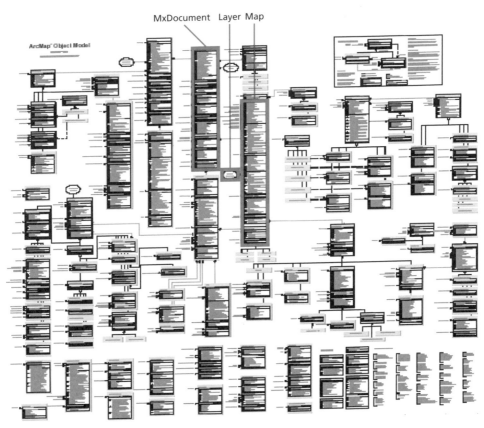

MxDocument   Layer   Map

ArcMap™ Object Model

Since your goal is to turn layers off, you need to find a path from the MxDocument class to the Layer class. To do this, you will zoom in on the diagram.

2 In Acrobat Reader, zoom in on the Map class, near the Layer in Map Layer oval, to match the graphic below. (The Zoom In tool looks like a magnifying glass. The oval is just above the exact center of the page.)

The Map class is connected to the Layer class by a gray line with a diamond and a star. The diamond tells you maps have layers; the star says a map can have many layers.

The Layer class is not shown on this diagram. Instead, it appears inside an oval (also known as a wormhole) with the name of the diagram, Map Layer, that includes it.

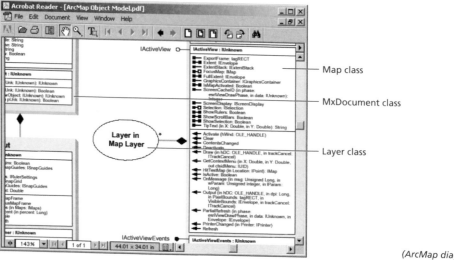

(ArcMap diagram)

3 In Acrobat Reader, open **Map Layer Object Model.pdf** in the **C:\ArcObjects \Diagrams** folder.

Layer gets a diagram to itself because it has so many coclasses.

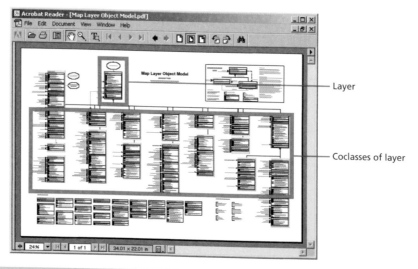

Next you'll take a closer look at the class properties to see how you can hop from one class to another.

4 Start ArcMap and open **ex11a.mxd** in the **C:\ArcObjects\Chapter11** folder.

5 When prompted for a password, enter **Carter**.

The ArcMap window maximizes and you see the streets of Washington, D.C., and several crime layers. On the Crime Analysis toolbar, a UIButton called ClearCrime has been added for you. This is the button that will turn off all the layers.

6 On the Crime Analysis toolbar, right-click ClearCrime and click View Source.

Visual Basic Editor opens and you see the ClearCrime click event procedure. It is empty at the moment.

To start writing code, you need to locate a property on MxDocument that gets an object of the Map class. The FocusMap property returns the IMap interface for the active map. (Remember, when we're using ArcObjects, a map is a data frame.)

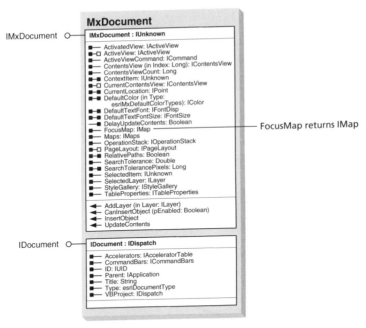

Since the FocusMap property is on the IMxDocument interface and the predefined ThisDocument variable points to the IDocument interface, you have to switch interfaces.

7 In the ClearCrime_Click event procedure, add the following two lines of code.

```
Dim pMxDoc As IMxDocument
Set pMxDoc = ThisDocument
```

Now that you have a variable pointing to the IMxDocument interface, you can get a map object in two lines of code, by declaring a variable and setting it. You declare the variable as IMap, because the FocusMap property returns the IMap interface.

8 Add two lines of code to get the active map from the map document.

```
Dim pMap As IMap
Set pMap = pMxDoc.FocusMap
```

Once you have a map object, you can get layers from that map with the Layer property, which is on the IMap interface of the Map class. The Layer property has an argument that specifies the layer's position in the table of contents. The top layer is at position 0, the second layer is at position 1, and so on.

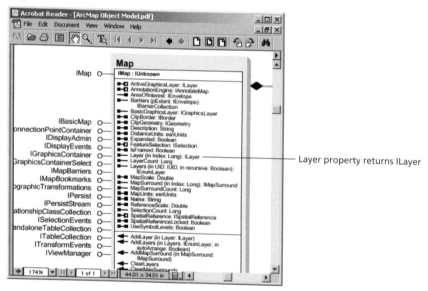

9 Declare an ILayer variable.

You declare the variable as ILayer because the Layer property returns the ILayer interface.

```
Dim pLayer As ILayer
```

You will set the pLayer variable inside a For Each loop. Each time the loop runs, the variable will be reset to a different layer, until all layers are turned off.

10 Start a For loop that begins with 0 and ends with the position number of the map's last layer.

```
For i = 0 to pMap.LayerCount - 1

Next i
```

The formula pMap.LayerCount – 1 gives you the index position of the last layer, no matter how many layers a map contains. For example, in a map with ten layers, the first layer is at position 0 (the first item in most ArcObjects lists is number zero) and the last is at position 9. LayerCount – 1 equals 9.

11 Inside the loop (before Next i), add a line of code to set the layer variable.

```
Set pLayer = pMap.Layer(i)
```

The layer's index number is stored in the variable i, which contains a 0 the first time through the loop. That will get the first layer on the map. The value of i will increment by 1 until there are no more layers in the map.

On the Layer diagram below, you can see that the ILayer interface has a Visible property. You turn layers on and off by setting the Visible property to true or false.

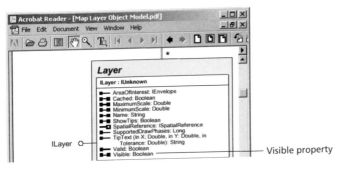

12 Add another line of code to turn the layer off.

```
pLayer.Visible = False
```

Getting layers

Although your code affects both the ArcMap table of contents and the map display, these parts of the application do not change unless explicitly told to do so. You will add some code to redraw these areas.

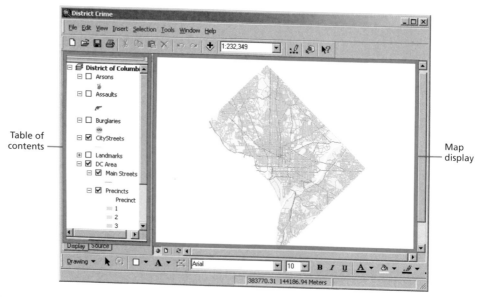

Table of contents

Map display

13 Outside the loop, after the Next i line and before the End Sub line, add the following line of code.

```
pMxDoc.UpdateContents
```

The UpdateContents method on the IMxDocument interface redraws the table of contents, but not the map display.

To redraw the map display, you need the IActiveView interface on the Map class. You can get there using the ActiveView property on IMxDocument.

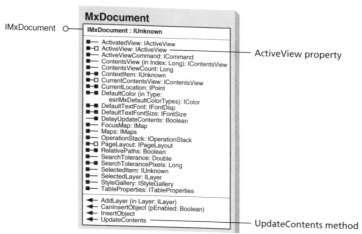

IMxDocument

ActiveView property

UpdateContents method

Once you get to IActiveView, you run its Refresh method to redraw the map display. (Since the IActiveView interface is found on both the Map and the PageLayout classes as shown below, the ClearCrime button will work whether the user is in data view or layout view. In one case, the map display redraws; in the other, the virtual page redraws.)

14 Declare and set a variable for the map display.

```
Dim pActiveView As IActiveView
Set pActiveView = pMxDoc.ActiveView
```

You can do this because the pMxDoc variable points to the IMxDocument interface, which has the ActiveView property.

15 Add a line of code to redraw the map display.

```
pActiveView.Refresh
```

After this line of code runs, all layers disappear from the map display.

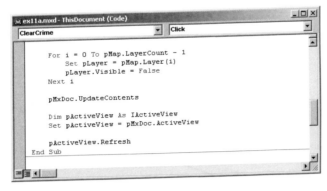

The code is ready to test.

16 Close Visual Basic Editor.

17 Turn all layers on by checking their boxes.

18 Click the ClearCrime button.

All the layers turn off.

19 Turn all the layers back on by checking their boxes.

Now that you have a button to turn off all the layers, you no longer need the table of contents for this purpose. You'll get rid of the table of contents and test the button again.

20 Click the Window menu and click Table of Contents.

Turning off the table of contents leaves you with more viewing space. If you had a button to turn the layers back on, you could use the application without the TOC.

21 Click the ClearCrime button.

All the layers turn off again.

22 Turn the table of contents back on and turn all the layers back on.

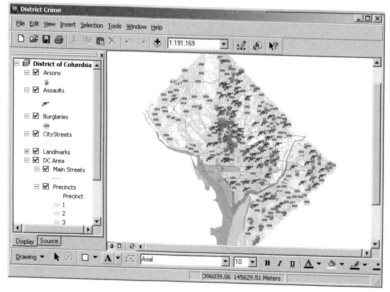

As an experiment, go back to the code and comment out the line that runs the UpdateContents method. Then click the ClearCrime button again to see that the table of contents check boxes remain checked. Click the Source and Display tabs at the bottom of the table of contents to force it to redraw. Go back and uncomment the UpdateContents line of code.

As another experiment, make a second button called AllCrime. Copy all the code from ClearCrime's click event to AllCrime's click event. Find the line of code in the For Each loop that sets layer visibility to False and change it to True. Try using your two new buttons together.

23 If you want to save your work, click the File menu in ArcMap and click Save As. Navigate to **C:\ArcObjects\Chapter11**. Rename the file **my_ex11a.mxd** and click Save. If you are continuing with the next exercise, leave ArcMap open. Otherwise close it.

Creating and assigning colors

Color is an integral part of making a map. You assign color to a layer's features, to text and graphics on a layout's page, and to the layout page itself. In ArcObjects, each color is an object that you create and then set properties for. The simplified diagram below shows the Color abstract class and five coclasses.

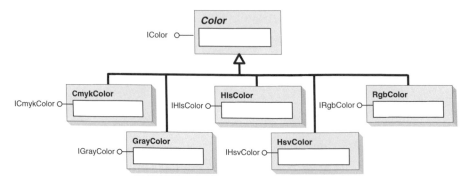

Each coclass represents a color model. Computer monitors display color with red, green, and blue light, so if you are making maps for the Internet, you would make colors with the RgbColor coclass. If your maps will be printed on paper, you would make colors with the CmykColor coclass because CMYK (cyan, magenta, yellow, black) is a color model standard in the print industry.

Each color coclass has properties that you set to mix and make a color. The RgbColor coclass has red, green, and blue properties.

The values for each property range from 0 to 255. (Why this range? Because a computer stores 256 values in one byte of memory.)

To make a sandy yellow color with the ArcMap user interface, you would set slider bars (or type in values) as shown below.

To make the same color with code, you set the properties of an RgbColor object:

```
pRgbColor.Red = 255
pRgbColor.Green = 255
pRgbColor.Blue = 190
```

In this exercise, you will change the background color of a layout page. By default this color is white, but you can change it to anything you like.

The layout page color has been changed from white to blue.

As in the previous exercise, you have to navigate to the object you want to work with—in this case, a color object on the layout page.

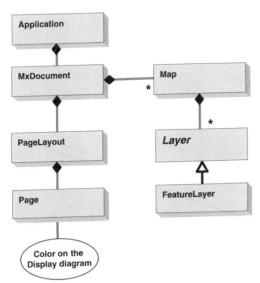

As before, you'll start with the MxDocument class, but this time you'll hop to Page Layout, and then to Page. At that point, you'll need to look at another diagram to see the Color classes.

Creating and assigning colors

When you get to the RgbColor class, you'll create a new color and set it equal to the Page's background property. After that, you'll refresh the display as you did in the last exercise.

Exercise 11b

You work for an adventure recreation company that organizes vacations to interesting places for windsurfing, desert survival hikes, and other rugged activities. For each trip that gets organized, you make the supporting maps. You want the background color of your map to reflect the theme of the vacation. If it's near water, the page color should be blue; if it's a desert tour, the page color should be sandy brown.

Before writing any code, you will examine the ArcMap object model diagram to determine which objects to use.

1 If you need to, start Adobe Acrobat Reader and open **ArcMap Object Model.pdf** from the **C:\ArcObjects\Diagrams** folder.

To change the page color, you need to find a path from MxDocument to Page. On the large diagram below, MxDocument is connected to PageLayout, and PageLayout is connected to Page. So in two lines of code you can hop from MxDocument to PageLayout, and in two more lines you can hop from PageLayout to Page.

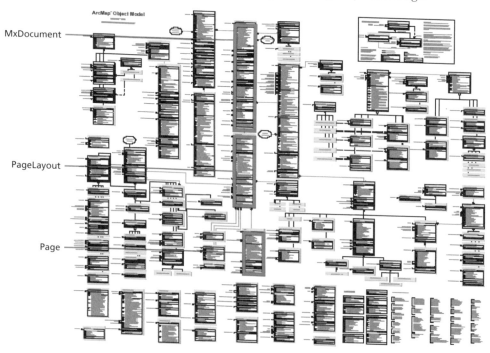

2 Zoom in on the Page coclass. It's near the bottom-center of the diagram.

The Page class has the BackgroundColor property, which is what you want to set. Since the BackgroundColor property requires the IColor interface, you need to get an object that has this interface.

IPage O—

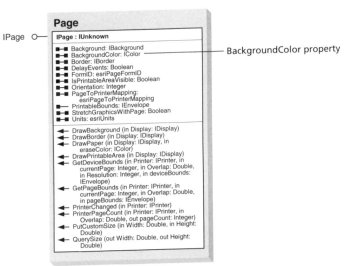

BackgroundColor property

In theory, you should see a line connecting the Page class to a class that implements IColor. But you don't, because on crowded diagrams, like this one, some connected classes can't be shown. To find classes that implement IColor, you might highlight IColor in the code window and press F1 to open the IColor page in the ArcObjects developer help, shown below.

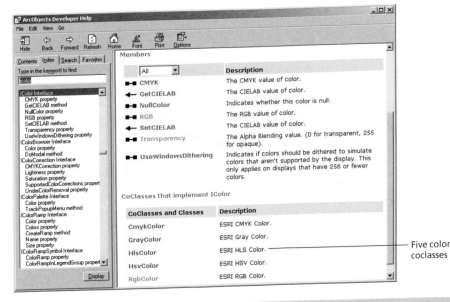

Five color coclasses

Creating and assigning colors

The help page shows five color coclasses. You can click each of them to see which interfaces they support and lists of properties and methods. But what if you want to see these color classes on an object model diagram?

Locating a class on one of the twenty or so object model diagrams can be like finding a needle in a haystack. Fortunately, there is an easy search method.

3 In Acrobat Reader, open **AllOMDs.pdf** in the **C:\ArcObjects\Diagrams** folder.

In the AllOMDs.pdf file, each diagram is bookmarked in the list at the left.

Bookmarks for each diagram —

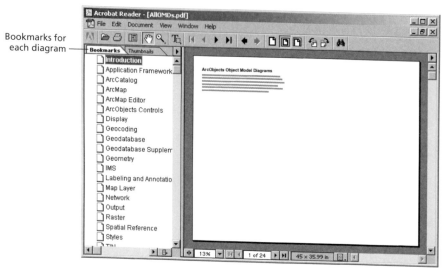

The Find button (called the Search button in Adobe Reader® 6.0) will find any word on any diagram and highlight it in blue. Before doing any finding, you will need to zoom in far enough to read the words.

4 In Acrobat Reader, click the View Menu and click Actual Size.

5 Click the Find button (the binoculars icon at the top). In the Find dialog box, type **RgbColor** and click Find.

The Acrobat Reader display area centers on the RgbColor coclass, which is also highlighted for you.

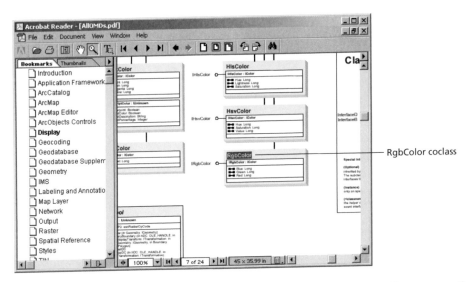

RgbColor coclass

6 Use the Hand tool (it looks and works like the ArcMap Pan tool) to move the RgbColor coclass to the lower right corner of the display, so you can see all the Color classes.

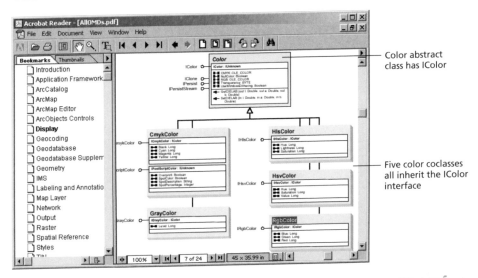

Color abstract class has IColor

Five color coclasses all inherit the IColor interface

You now know which classes (MxDocument, PageLayout, Page, and RgbColor) are needed to get the layout page and set its color. Next you will write the code to set a new page color.

7 Start ArcMap and open **ex11b.mxd** in the **C:\ArcObjects\Chapter11** folder.

When the map opens you see potential vacation sites in the country of Belize.

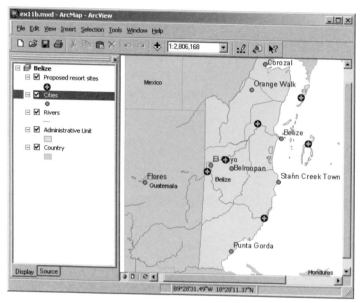

8 Switch from Data View to Layout View.

On the Layout toolbar, a new menu called Page Color has been added for you. It contains one choice, a UIButton called BluePage. You will code this button to set the background color of your layout page.

9 Click the Tools menu and click Customize.

10 On the Layout toolbar, click Page Color, right-click on BluePage, and click View Source.

You see an empty click event procedure.

11 Add the following code to get to IMxDocument.

```
Dim pMxDoc As IMxDocument
Set pMxDoc = ThisDocument
```

12 Add two more lines to get the page layout.

```
Dim pPageLayout As IPageLayout
Set pPageLayout = pMxDoc.PageLayout
```

13 Add two more lines to get the page.

```
Dim pPage As IPage
Set pPage = pPageLayout.Page
```

To make a color object, you use the Set and New keywords.

14 Create an RgbColor object to serve as the background color of the page.

```
Dim pRgbColor As IRgbColor
Set pRgbColor = New RgbColor
```

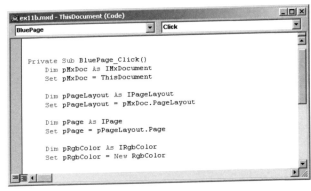

15 Set the Red, Green, and Blue properties to get the desired shade of blue.

```
pRgbColor.Red = 100
pRgbColor.Green = 150
pRgbColor.Blue = 255
```

The color properties you just set are found on the IRgbColor interface. But recall from step 2 that the BackgroundColor property returns the IColor interface. This means you must use QueryInterface to switch from IRgbColor to IColor. (The principle of inheritance tells you that you can do this. The RgbColor coclass implements all the interfaces shown on the IColor abstract class.)

16 Declare an IColor variable and set it equal to the variable that is already pointing to the RgbColor object.

```
Dim pColor As IColor
Set pColor = pRgbColor
```

17 Assign the color to the page's BackgroundColor property.

```
pPage.BackgroundColor = pColor
```

Creating and assigning colors

18 Get and refresh the page layout's display. (As with the map display, you use the IActiveView interface to do this.)

```
Dim pActiveView As IActiveView
Set pActiveView = pPageLayout

pActiveView.Refresh
```

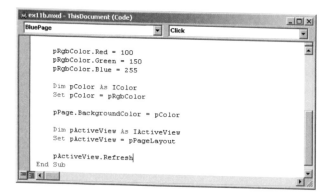

The code is ready to test.

19 Close Visual Basic Editor.

20 In ArcMap, on the Layout toolbar, click Page Color and click BluePage.

The page color changes to blue.

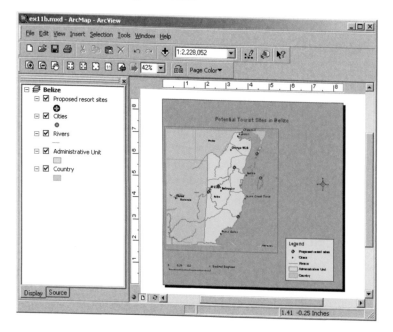

To change the color back to white, you could make a second UIButton called WhitePage, add it to the Page Color menu, and copy all the code from BluePage's click event to WhitePage's click event. Then you would edit the property values for pRgbColor as shown:

```
pColor.Red = 255
pColor.Green = 255
pColor.Blue = 255
```

For desert area maps where you want a sand color, you could make a third UIButton called SandPage, add it to the menu, copy the code, and edit the color values as shown:

```
pColor.Red = 215
pColor.Green = 194
pColor.Blue = 158
```

21 If you want to save your work, click the File menu in ArcMap and click Save As. Navigate to **C:\ArcObjects\Chapter11**. Rename the file **my_ex11b.mxd** and click Save. If you are continuing with the next chapter, leave ArcMap open. Otherwise close it.

Making tools

Reporting coordinates
Drawing graphics
Using TypeOf statements

Back in chapter 2, you learned about making commands when you created the TaxCalculator UIButton. You may be wondering why you are just now learning how to make a UITool. It's because coding a tool requires working with a variety of objects, including maps, layers, and geometry objects like points. You wouldn't have been ready without first getting some experience in working with classes and interfaces, and reading object model diagrams.

When you coded the TaxCalculator UIButton, you put the code in the button's click event. When you write code for a tool, there's more to it, because the user interacts with the tool. They might pan the map extent, draw graphics, select features, or drag a zoom rectangle. In all these situations, the user moves the mouse pointer around the map, sometimes clicking or double-clicking on a location.

For you, the programmer, this means that a tool has more events to code than a button. Instead of click events, tools have MouseDown, MouseUp, and MouseMove events that run when the user interacts with the map display. They have CursorID events that define the appearance of the mouse pointer in the display. (For example, the Pan tool cursor looks like a hand and the Measure tool cursor looks like a ruler.) They also have Enabled events that can be used to disable the tool under specific conditions.

In this chapter, you will create a UITool and code some of its events. Your tool will report coordinates of the cursor's location and draw graphic points wherever the user clicks on the map display. If the map is in layout view, the tool will be grayed out.

Reporting coordinates

Code in a tool's MouseMove event procedure runs whenever the user, with the tool selected, moves the mouse pointer in the map display. As shown below, the procedure has four variables in its argument list.

```
Private Sub UIToolControl1_MouseMove ( _
    button As Long, shift As Long, x As Long, y As Long)

End Sub
```

Oddly enough, it's the user who sets these variables. The x and y variables represent the location of the mouse pointer on the map display. Since this is the MouseMove event, as the user moves the mouse the values of x and y change. The variable values are in the pixel units of the user's computer's monitor.

The button and shift variables let you code alternative situations. For example, you could write an If Then statement that runs one block of code if the user is holding down the left mouse button and a different block if they are holding down the right button.

The button variable has a value of 1 or 2. The value is 1 when the user is holding down the left mouse button (while moving the mouse); the value is 2 when they are holding down the right button. The shift variable has a value of 0 or 1. The value is 0 when the Shift key is not depressed; the value is 1 when it is depressed.

The MouseDown and MouseUp events have the same four variables as the MouseMove event. Code in a tool's MouseDown and MouseUp event procedures runs when the user, with the tool selected, clicks in the map display or releases a mouse button.

Suppose you wanted to make a tool that used a message box to display the coordinates of a location the user clicked on. You might code the tool's MouseDown event procedure like this:

```
Private Sub UIToolControl1_MouseDown ( _
    button As Long, shift As Long, x As Long, y As Long)

    MsgBox "X: " & x & ", Y: " & y

End Sub
```

A user clicking on Buenos Aires would get a result like the one in the following graphic.

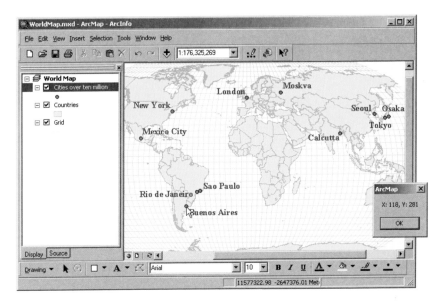

That's useful if the user wants to know how many pixels Buenos Aires is from the edge of the display. To get meaningful geographic units, however, will take some ArcObjects programming.

The IMxDocument interface has a property called CurrentLocation that gets the location of the mouse pointer in map units (meters or feet for projected data; latitude–longitude for unprojected data).

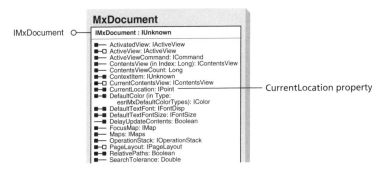

Since the CurrentLocation property requires the IPoint interface, there should be a line connecting MxDocument to a class that implements IPoint. There isn't, however, because MxDocument is connected to more than thirty classes and the diagram isn't big enough to show them all.

Instead, you can look up the IPoint interface in the developer help.

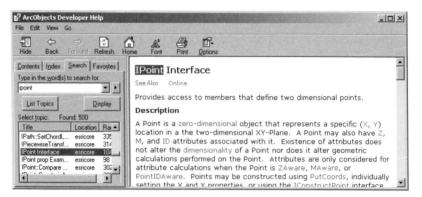

If you scroll to the bottom of the topic, you'll see a list of the coclasses that implement IPoint.

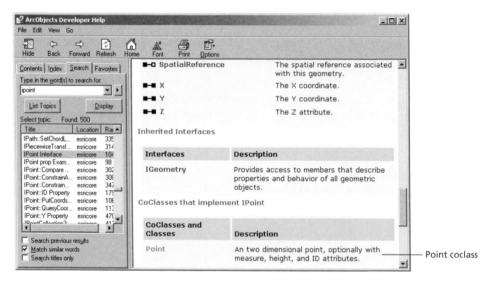

Point coclass

The only coclass that implements IPoint is Point. The Point coclass, shown below, has x and y properties that store coordinates in map units. The double barbells tell you that you can both get and set these properties.

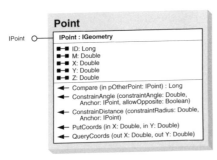

Point

IPoint : IGeometry
■–■ ID: Long
■–■ M: Double
■–■ X: Double
■–■ Y: Double
■–■ Z: Double
◄– Compare (in pOtherPoint: IPoint) : Long
◄– ConstrainAngle (constraintAngle: Double, Anchor: IPoint, allowOpposite: Boolean)
◄– ConstrainDistance (constraintRadius: Double, Anchor: IPoint)
◄– PutCoords (in X: Double, in Y: Double)
◄– QueryCoords (out X: Double, out Y: Double)

IPoint ○

To get the mouse pointer's current location in map units, you navigate from MxDocument to Point. Then you use object.property syntax to get the point's x and y values. Starting from the predefined ThisDocument variable, your MouseDown event procedure would look like this:

```
Dim pMxDoc As IMxDocument
Set pMxDoc = ThisDocument

Dim pPoint As IPoint
Set pPoint = pMxDoc.CurrentLocation

MsgBox "Longitude: " & pPoint.X & _
       ", Latitude: " & pPoint.Y
```

Now, when the user clicks on Buenos Aires, they see a message like the one below.

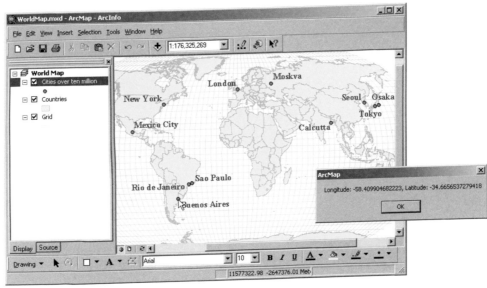

If you were using projected data, you would change the message box text accordingly.

Exercise 12a

You work as a GIS specialist for the Grand Canyon search and rescue team. The canyon's rugged terrain can make it hard to find hikers in distress and just as hard to airlift them out. If an airlift isn't feasible, their location is captured with a GPS unit on board a helicopter. The latitude–longitude coordinates are then sent to a dispatcher, who pinpoints the location on a paper map, decides which rescuers to send, and directs the team to the scene.

The dispatcher and the rescue team describe their positions in various ways, from the familiar "We are one kilometer south of the bridge," to reading off the GPS unit "Latitude 36, longitude –112." It would be useful to have an application that could report locations in both meters and latitude–longitude.

In this exercise, you'll make a RescueSite tool that displays latitude–longitude coordinates in ArcMap. You'll add a new tool to a toolbar and choose a cursor for it. Since the dispatchers need to see coordinates while moving the tool around the map, you'll write the code in the MouseMove event and display the values in the ArcMap status bar.

1 Start ArcMap and open **ex12a.mxd** in the **C:\ArcObjects\Chapter12** folder.

When the map opens, you see an area of the Grand Canyon centered on Hopi Point.

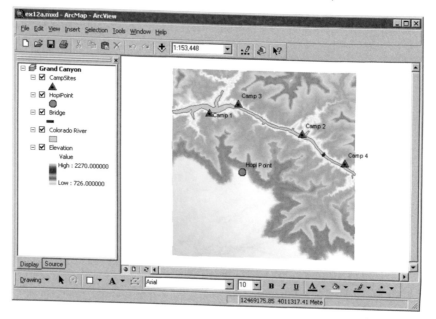

Although the source data is unprojected, the data frame's display units have been set to meters, so these are the values that you see in the status bar. You won't replace these units, but you'll add a second display of lat–long coordinates.

Your code will display
lat–long units here

Location coordinates in
meters display here

2 Click the Tools menu and click Customize. Click the Commands tab. At the bottom Categories list, click [UIControls].

By default, your customizations will be saved to the current map document (ex12a.mxd).

3 Click the New UIControl button. For the UIControl type, click UIToolControl.

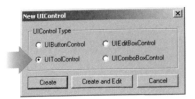

4 Click Create.

In the Commands list, you see a new tool named Project.UIToolControl1.

5 Click on Project.UIToolControl1 and replace the text with **Project.RescueSite**.

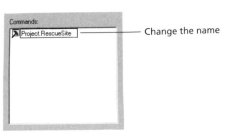

Change the name

6 Drag Project.RescueSite to the bottom of the Tools toolbar as shown in the following graphic. (If your toolbar is horizontal, add the new tool at the end.)

— Add tool

Next you will change the tool's icon.

7 Right-click on the new tool, point to Change Button Image, and click Browse.

8 In the Open file browser, navigate to **C:\ArcObjects\Data** and click **Helicopter.bmp**. Click Open.

You see a helicopter icon on the tool.

Now you will code the tool's CursorID event procedure. This defines the appearance of the cursor when the tool is used in the map display.

9 Right-click the RescueSite tool and click View Source.

Visual Basic Editor opens with the ThisDocument code module open. Code for all UIControls is stored in the ThisDocument module.

10 With RescueSite selected in the object list, click the procedure list drop-down arrow and click CursorID.

You see the CursorID event wrapper lines. Before going any further, you'll get some online help.

11 On any blank line in the Code window, type in **SetCursor**. Use your mouse to highlight **SetCursor**.

12 Press the F1 key. In the Context Help window, with the esri Framework library selected, click Help. Scroll down to the description and pictures of some cursor IDs.

F1 searches for the highlighted term in the help system. If the term is found, its help page opens. If it's an ArcObjects term, the ArcObjects developer help opens. If it's a VBA term, the Visual Basic help opens.

The SetCursor help page shows the available icons and their ID numbers. You will use cursor 3, which looks like a crosshair.

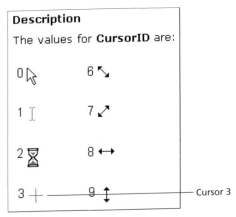

— Cursor 3

If you scroll down the help page, you will find code for using cursor icons other than the ten shown here.

13 Close the help window. Press Delete to remove SetCursor from the Code window.

Reporting coordinates

14 In the CursorID event, add the following line of code to set the cursor ID to icon number 3.

```
RescueSite_CursorID = 3
```

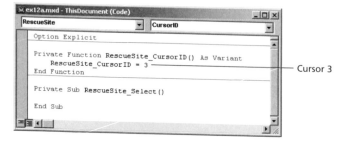

Cursor 3

15 Close Visual Basic Editor.

Visual Basic Editor must be closed for a tool's CursorID event to run. Now you will test the cursor.

16 Click the RescueSite (Helicopter) tool and move the cursor over the map. It changes to a crosshair.

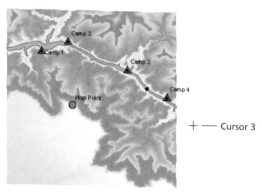

Cursor 3

The ArcMap status bar displays meters in message pane 2. You'll add lat–long values in message pane 0.

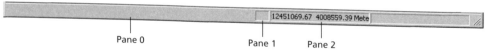

Pane 0   Pane 1   Pane 2

17 Right-click the RescueSite tool and click View Source to open the ThisDocument code module.

18 With RescueSite selected in the object list, click the procedure list drop-down arrow and click MouseMove.

19 In the MouseMove event procedure, add the following code to declare and set a map document variable that points to IMxDocument.

```
Dim pMxDoc As IMxDocument
Set pMxDoc = ThisDocument
```

20 Add two lines to declare a point variable and set it with IMxDocument's CurrentLocation property.

```
Dim pPoint As IPoint
Set pPoint = pMxDoc.CurrentLocation
```

This code stores the current location of the cursor in the pPoint variable. Now you need to get this information into the ArcMap status bar.

The StatusBar class has a Message property on its IStatusBar interface for displaying text strings. To get the status bar, you use the StatusBar property on IApplication. You may recall from chapter 10 that the Application class has a predefined object variable called Application. That will be your starting point.

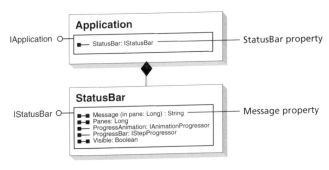

21 Add two lines of code to declare and set a status bar variable.

```
Dim pStatus As IStatusBar
Set pStatus = Application.StatusBar
```

Now that you have the status bar, you can set its Message property. The Message property has an argument for specifying which pane, or section, of the status bar to use. (The ArcMap status bar has four panes; you will use the left-most pane, which is identified as 0.)

Reporting coordinates

22 Add the following line of code to report latitude and longitude in the status bar's first message pane. Use an underscore for line continuation. Use & to concatenate text and location values.

```
pStatus.Message(0) = _
    "Latitude: " & pPoint.y & ", Longitude: " & pPoint.x
```

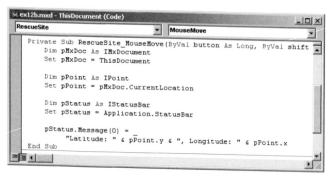

23 Close Visual Basic Editor.

24 If necessary, click the RescueSite tool in ArcMap to select it. Move the mouse over Hopi Point.

You don't have to click to see the coordinates because your code is in the MouseMove event procedure.

Your code displays
lat–long units here

Hopi Point's latitude and longitude (about 36, –112) appear in the first message pane in the status bar. As you move the mouse, the coordinates change.

25 If you want to save your work, click the File menu in ArcMap and click Save As. Navigate to **C:\ArcObjects\Chapter12**. Rename the file **my_ex12a.mxd** and click Save. If you are continuing with the next exercise, leave ArcMap open. Otherwise close it.

Drawing graphics

To make a direction map, you might use a streets layer for reference and then draw graphics on it, like the circles, red line, and text labels on the map below.

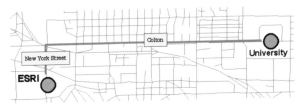

You can also add graphics to layouts, like the title and directions in the next example. The main difference between graphics on a map and graphics on a layout are their coordinates. Map graphics have coordinates in map units like feet or meters, while layout graphics have page coordinates like inches or centimeters.

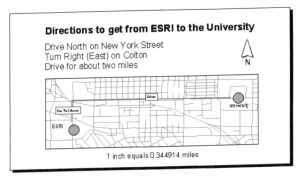

You've probably used graphics on many occasions. You know how to make them and set their properties from the user interface. One of the challenges, and part of the fun, of ArcObjects programming is finding familiar objects on a diagram and figuring out how to make them, set their properties, and run their methods by writing code.

Graphics belong to an abstract class called Element. As shown on the diagram below, both Maps and PageLayouts are composed of Elements. Each has an IGraphicsContainer interface with methods, like AddElement, to add, delete, and move its elements.

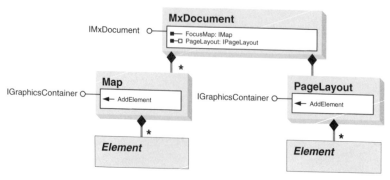

The next diagram shows that the Element class has two abstract subclasses: GraphicElement and FrameElement. To make graphics on a map, you create objects out of the coclasses under GraphicElement. These include LineElement, TextElement, and MarkerElement. (You'll learn more about the FrameElement class—which is used to make data frames, legends, north arrows, and scale bars—in chapter 19.)

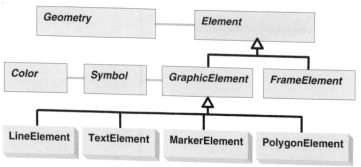

The diagram also shows that graphic elements have a symbol and that symbols have a color. You made colors in chapter 11 and you will make symbols in chapter 15. In this exercise, you'll use default values (for marker elements, the default is a red square) and not worry about making symbols and colors.

The last thing the diagram shows is that Element is connected to Geometry. This means that elements are associated with geometry objects, like points, lines, and polygons, as shown below.

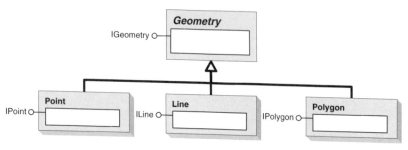

Geometry objects, then, are your starting point to create graphic elements. Geometry objects have to be compatible with their associated element. If you are making a polygon element, you would begin by making a polygon geometry.

Say you want to create a marker element to mark the city of London. The geometry for a marker element is a point, so you begin by creating a point with the code below.

```
Dim pPoint As IPoint
Set pPoint = New Point
```

To define the point's location, you set its x and y properties. The following code puts the point in London, at 0 degrees longitude and 53.5 degrees north latitude:

```
pPoint.X = 0
pPoint.Y = 53.5
```

Having created or gotten a point (in the last exercise you got a point with the CurrentLocation property), you make a marker element (the point's visual representation). The MarkerElement coclass is shown below.

The following code makes a new marker element:

```
Dim pMarkerElement As IMarkerElement
Set pMarkerElement = New MarkerElement
```

Next, you associate the point with the marker element. The abstract Element class has a Geometry property on its IElement interface to do this.

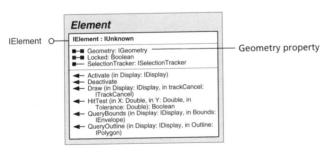

Since the new marker element was declared to the IMarkerElement interface, you would switch interfaces to IElement. Then you use the point to set IElement's Geometry property.

```
Dim pElement As IElement
Set pElement = pMarkerElement

pElement.Geometry = pPoint
```

Once you have made the graphic, you still have to store it in the map's graphics container. The graphics container is the collection of all graphics in the map. Shown below, the Map class's IGraphicsContainer interface has the AddElement method to add graphics to the map.

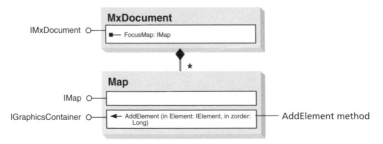

To get the graphics container, you use the IMxDocument's FocusMap property, which returns the map's IMap interface.

```
Dim pMap As IMap
Set pMap = pMxDoc.FocusMap
```

FocusMap returns IMap, so you switch interfaces to the map's IGraphicsContainer interface.

```
Dim pGraphics As IGraphicsContainer
Set pGraphics = pMap
```

You are probably getting familiar with using QueryInterface. You use a property that returns an interface you don't need, and you switch interfaces to get the interface you do need. It always takes four lines of code: two to declare and set the interface variable that you don't need and two more to declare and set the interface variable that you want.

You are now ready to learn a shortcut. VBA knows how to do the QueryInterface for you. Any time a property returns an object's interface, you can declare your variable to any interface of that object and VBA will get it for you.

In the example above, FocusMap returns a map's IMap interface. You have no use for IMap because you want to work with graphics and the map's IGraphicsContainer interface. You don't need to declare an IMap variable. You can declare the IGraphicsContainer variable and set it with the FocusMap property, as shown below.

```
Dim pGraphics As IGraphicsContainer
Set pGraphics = pMxDoc.FocusMap
```

FocusMap returns IMap. VBA sees, however, that you want IGraphicsContainer (you declared a variable for it), so it does the QueryInterface for you and sets the variable correctly. The shortcut saves you two lines of code. The four-line QI technique is fine, especially when you are not used to switching interfaces, but if you don't need an IMap variable, you shouldn't always have to write two lines of code to make one.

Whichever method you use to get the graphics container, the next thing to do is to store the graphic in it. To do that, you run the IGraphicsContainer interface's AddElement method.

```
pGraphics.AddElement pElement, 0
```

After storing the graphic, you refresh the display. You did this in the last chapter after writing code to turn off all the layers in the table of contents. In this situation, however, where you're just adding a graphic, it's not efficient to redraw the entire map display.

The PartialRefresh method, located on the Map class's IActiveView interface shown below, saves you from having to refresh the entire display after minor changes. It has arguments to control which caches are refreshed (phase), which individual layers or elements are refreshed (data), and which parts of the screen are refreshed (envelope).

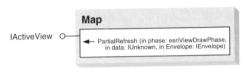

A cache is sort of like a screen capture or photo image of what's displayed in a map's active view area. ArcMap uses caches to redraw itself more quickly.

Say, for example, that you cover the ArcMap window with another window (like the VBA window). When ArcMap comes forward again, everything in the active view must be redrawn, even though nothing has changed. Now say that the map contains several layers that display hundreds of features. It's not efficient for ArcMap to go to each layer, locate its source data, and redraw each feature's geometry. The cache remembers what the data looks like, and can draw it fast since it is only an image.

At any given time, maps have several caches: one for graphics, one for layers, and one for selected features. The Refresh method that you have been using redraws every cache. The PartialRefresh method, on the other hand, can redraw just one cache. Suppose that you have code that changes the selected set of features. PartialRefresh can redraw just the cache with the selected features, saving the time of redrawing the graphics and layers caches.

PartialRefresh's data and envelope arguments help make redraws even faster. With the data argument, you can specify a particular layer or graphic to redraw. In this exercise, you are creating a tool to draw a graphic point. You will use PartialRefresh to redraw that single graphic on the graphics cache with code like the line below.

```
pActiveView.PartialRefresh _
    esriViewGraphics, pElement, Nothing
```

Instead of redrawing the entire active view area, PartialRefresh's envelope argument lets you specify a rectangle to redraw. For example, if your code adds a cluster of five polygon graphics, you can get their combined extent rectangle, or envelope, and redraw just that area. If you don't want to use the data or envelope arguments, you use the Nothing keyword.

In this exercise, you will add code to the RescueSite tool. The code will let the user draw markers on the map and refresh the display for each one.

Exercise 12b

As dispatchers identify a location with the RescueSite tool, coordinates appear in the status bar. Besides seeing the rescue site's coordinates, dispatchers want to be able to mark the site. In this exercise, you will add code to the RescueSite tool's MouseDown event to draw graphic points on the map.

1 Start ArcMap and open **ex12b.mxd** in the **C:\ArcObjects\Chapter12** folder.

When the map opens, you see the Grand Canyon and the RescueSite tool on the Tools toolbar.

2 Right-click the RescueSite tool and click View Source.

In the ThisDocument code module, you see the code for the CursorID and MouseMove event procedures you wrote in the last exercise.

3 With RescueSite selected in the object list, click the procedure list drop-down arrow and click MouseDown.

4 Add four lines to declare and set variables for the map document and a point.

```
Dim pMxDoc As IMxDocument
Set pMxDoc = ThisDocument
Dim pPoint As IPoint
Set pPoint = pMxDoc.CurrentLocation
```

This is the same code you used in the last exercise to get the coordinates of the point at the current mouse location. This time, you will create a marker element and assign it the geometry of this point. As shown below, IElement has the Geometry property you need.

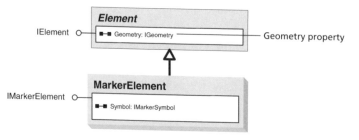

It might look as if you have to declare a variable to IMarkerElement and then switch to the IElement interface. Because of class inheritance, however, you can declare the variable directly to IElement. (If you were going to set your own symbology, instead of using the default, you would declare the variable to IMarkerElement, because it has the Symbol property.)

5 Declare a variable and create a new marker element.

```
Dim pElement As IElement
Set pElement = New MarkerElement
```

6 Set the marker element's Geometry property equal to the point returned by the CurrentLocation property.

```
pElement.Geometry = pPoint
```

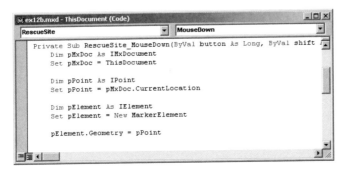

```
ex12b.mxd - ThisDocument (Code)
RescueSite                              MouseDown

Private Sub RescueSite_MouseDown(ByVal button As Long, ByVal shift
    Dim pMxDoc As IMxDocument
    Set pMxDoc = ThisDocument

    Dim pPoint As IPoint
    Set pPoint = pMxDoc.CurrentLocation

    Dim pElement As IElement
    Set pElement = New MarkerElement

    pElement.Geometry = pPoint
```

Now you want to get the map's graphics container. Here is your chance to use the QI shortcut. To get a map, you usually would use the FocusMap property on IMxDocument. FocusMap returns the IMap interface. Then you switch interfaces to IGraphicsContainer.

This time, however, you will declare an IGraphicsContainer variable and set it with the FocusMap property and let VBA take care of the QueryInterface for you.

7 Declare and set an IGraphicsContainer variable.

```
Dim pGraphics As IGraphicsContainer
Set pGraphics = pMxDoc.FocusMap
```

You store the element in the map's graphics container with the AddElement method.

AddElement has two arguments: an IElement object (which you just made in step 5) and an index number, which defines the element's ordered position in the graphics container. The 0 position is in the front. (It's the position a selected graphic has when you click Bring to Front.)

8 Write a line of code to add the element to the map's graphics container.

```
pGraphics.AddElement pElement, 0
```

So far you have made a marker element, associated it with point geometry, and added it to the map's graphics container. The last thing to do is refresh the display.

You want the user to see each graphic they add, so you'll refresh the display every time the AddElement method runs. Instead of refreshing the entire display area, you will use the PartialRefresh method's second argument, to refresh just the new graphic.

Since the PartialRefresh method is on the map's IActiveView interface, you need to declare and set an IActiveView variable. The pGraphics variable (from step 7) points to the map's IGraphicsContainer interface, so you can use it to QueryInterface.

9 Declare and set an IActiveView variable.

```
Dim pActiveView As IActiveView
Set pActiveView = pGraphics
```

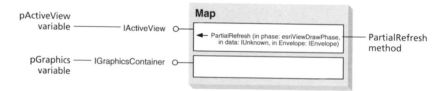

10 Add a line of code to run the PartialRefresh method to refresh the new element on the graphics cache (without an envelope).

```
pActiveView.PartialRefresh _
    esriViewGraphics, pElement, Nothing
```

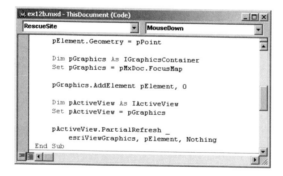

For the names of other draw phases (that control which caches are refreshed), see the topic *esriViewDrawPhase Constants* in the developer help.

The code is ready to test.

11 Close Visual Basic Editor.

12 Click the RescueSite tool and click a few times between Camp 2 and Camp 3.

With each click, a red square (the default marker symbol) draws on the map.

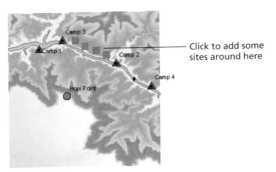

Click to add some sites around here

Early in this chapter, you learned that mouse events have button variables. You could make the RescueSite tool more useful by using the two mouse buttons for different tasks. For example, while the left button draws red squares for rescue sites, you could use the right button to draw blue circles for trail obstructions, like landslides and fallen trees.

Although you haven't yet learned how to set symbology, you can try out the code below to see mouse button variables in action. Add the following If Then statement to the MouseDown event just before the PartialRefresh method. The code checks to see which mouse button has been clicked (1 is left and 2 is right). If the left button is clicked, the code doesn't run and the default red square is drawn. If the right button is clicked, the code below runs and a blue circle is added to the map.

```
If button = 2 Then
    Dim pSymbol As IMarkerSymbol
    Set pSymbol = New SimpleMarkerSymbol

    Dim pColor As IRgbColor
    Set pColor = New RgbColor
    pColor.RGB = vbBlue

    pSymbol.Color = pColor

    Dim pMElement As IMarkerElement
    Set pMElement = pElement
    pMElement.Symbol = pSymbol
End If
```

The default symbol for a SimpleMarkerSymbol is a black circle. The code above changes the circle's color to blue.

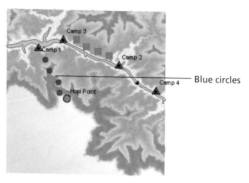

13 If you want to save your work, click the File menu in ArcMap and click Save As. Navigate to **C:\ArcObjects\Chapter12**. Rename the file **my_ex12b.mxd** and click Save. If you are continuing with the next exercise, leave ArcMap open. Otherwise close it.

Using TypeOf statements

In chapter 8, you learned about the Type Mismatch error message, which occurs when your code encounters an unexpected data type, such as a string where an integer is required.

Type mismatches can occur in lots of different situations. One example is when you are working with different layer types (like feature layers and raster layers) in the same map document. Suppose you have some code that displays the name of any layer a user selects. It looks like this:

```
Dim pMxDoc As IMxDocument
Set pMxDoc = ThisDocument

Dim pLayer As ILayer
Set pLayer = pMxDoc.SelectedLayer

MsgBox pLayer.Name
```

You decide to write some additional code that switches interfaces and reports the number of rows in the matrix of any selected raster layer.

```
Dim pRasterLayer As IRasterLayer
Set pRasterLayer = pLayer
MsgBox pRasterLayer.RowCount & " rows"
```

Everything works fine as long as the selected layer is in fact a raster layer. But if it's a feature layer, or some other kind of layer, you get a type mismatch.

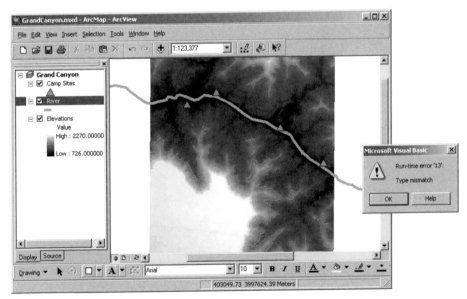

TypeOf statements are designed to help you avoid type mismatches and other runtime problems. They check to see whether an object variable points to a particular interface (or whether it *can* point to that interface through QueryInterface). If it does, the TypeOf statement returns true. If it doesn't, the statement returns false.

The syntax for a TypeOf statement looks like this:

```
TypeOf pLayer Is IRasterLayer
```

If pLayer is pointing to the interface of an object that also has the IRasterLayer interface, the statement returns true.

Since TypeOf returns true or false, it can be used in If Then statements like the one below. If the selected layer is a raster layer, you run the code you've already written. If it isn't, you display a message box that tells the user what to do.

```
If TypeOf pLayer Is IRasterLayer Then
    Dim pRasterLayer As IRasterLayer
    Set pRasterLayer = pLayer
    MsgBox pRasterLayer.RowCount & " rows"
Else
    MsgBox "Please select a raster layer"
End If
```

In the next exercise, you will use the TypeOf statement with the RescueSite tool to test whether the user is in data view or layout view.

The RescueSite tool draws graphics at locations specified in map units. Since layouts are in page units, like inches or centimeters, the RescueSite tool is of no use there and should be disabled (grayed out). You will write the code to do this in the tool's Enabled event procedure. This event procedure can be used to make any UIControl available or unavailable to the user.

Disabled   Enabled

Exercise 12c

The RescueSite tool assumes that the user is in data view. Once in a while, dispatchers try to use the tool in layout view by mistake. They get an error because the CurrentLocation property encounters page units instead of map units. (If you try it yourself, you'll find that clicking to dismiss the error message gets a new pair of page coordinates from the layout and causes another error. To get out of the loop, move the error message away from the layout.)

1 Start ArcMap and open **ex12c.mxd** in the **C:\ArcObjects\Chapter12** folder.

When the map opens, you see the Grand Canyon and the RescueSite tool.

2 Right-click the RescueSite tool and click View Source.

The ThisDocument code module contains the CursorID, MouseDown, and MouseMove event procedures that you have written in the last exercises. You will code a TypeOf statement in the Enabled event procedure.

3 With RescueSite selected in the object list, click the procedure list drop-down arrow and click Enabled.

You see the Enabled event procedure's wrapper lines added to the ThisDocument code module. You will write code to get the active view, so you can test it with the TypeOf statement.

4 Declare and set a variable to get to the IMxDocument interface.

```
Dim pMxDoc As IMxDocument
Set pMxDoc = ThisDocument
```

IMxDocument has the ActiveView property.

5 Get the active view.

```
Dim pActiveView As IActiveView
Set pActiveView = pMxDoc.ActiveView
```

This code gets the active view, whether it is layout view or data view. You will add a TypeOf statement to check which it is.

6 Begin an If Then statement. For its expression, use the TypeOf statement to see if pActiveView also has the IPageLayout interface.

```
If TypeOf pActiveView Is IPageLayout Then

End If
```

You are testing the pActiveView variable to see if it points to the interface of an object that also has the IPageLayout interface. If the statement returns true, it means that the object has IPageLayout and therefore that the user is in layout view.

7 Inside the If Then statement, indent and add a line of code to disable the tool.

```
RescueSite_Enabled = False
```

If you look at the Enabled event procedure's wrapper lines, you will see that it is defined as a function. As you know from chapter 6, functions return a value. When a control's Enabled event returns false, VBA grays out the control.

You will add an Else to the If Then statement so that you can add a line of code to enable the tool when the user is not in layout view.

8 Outdent and add the Else statement.

```
Else
```

9 Indent and add the following line to enable the tool if the user is in data view.

```
RescueSite_Enabled = True
```

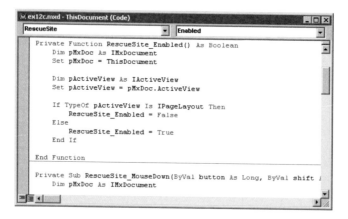

10 Close Visual Basic Editor.

11 Click the RescueSite tool and click the map a few times with both the left and right mouse buttons.

12 Click the View menu and click Layout View.

The RescueSite tool is grayed out. The tool and its code are protected from misuse.

Disabled

13 If you want to save your work, click the File menu in ArcMap and click Save As. Navigate to **C:\ArcObjects\Chapter12**. Rename the file **my_ex12c.mxd** and click Save. If you are continuing with the next chapter, leave ArcMap open. Otherwise close it.

Executing commands

Using CommandItems and CommandBars

ArcGIS toolbars are composed of commands. The UIControls that you create and drag to toolbars are commands, and so are all the tools, buttons, and menu choices that make up the ArcMap and ArcCatalog toolbars.

Every command has source code behind it. You can view the source code for UIControls that you make yourself, not for the controls that ESRI programmers have made for ArcMap and ArcCatalog. Even though the code for the predefined controls is not directly accessible to you (and is mostly written in C++, anyway), you can call it and tell it to run, almost as if you were calling a subroutine.

Commands have an interface called ICommandItem and this interface has an Execute method. When you run the method, the command executes—just as if you had clicked a button or made a menu choice on the user interface.

The ability to execute commands saves you from having to write new code for common operations. For example, instead of writing a procedure from scratch to print a map, you could write:

```
PrintButton.Execute
```

Or say that your users always finish making a map by clicking the same three buttons: Full Extent, Save, and Print. You could consolidate that process into a single button click by making a UIButton and writing three lines of code that run the Execute method on each button.

Toolbars, unlike commands, do not carry out operations, but sometimes you want to manipulate them in your code—for instance, to specify that a certain toolbar always opens with a map document or docks in a definite position. A lot of the code you write to work with toolbars is the same that you use to execute commands.

Using CommandItems and CommandBars

Say that you've written some code to zoom to a selected feature. When the code zooms in on a country—Bolivia, for example—the edges of the feature might touch corners of the data frame's neatline. In addition, you can't see the names of the neighboring countries.

Users compensate for these unwanted effects by clicking the Fixed Zoom Out button on the Tools toolbar.

Fixed Zoom Out

As a programmer, you would like to automate this process, so that whenever a user zooms to a feature, they then zoom out by one increment without having to click another button. You can do this by executing the Fixed Zoom Out command in your code.

On the diagram below, four classes are involved in the process. All commands, including the Fixed Zoom Out button, belong to the CommandItem class. You use the Execute method on the ICommandItem interface to tell a command to run its code.

Toolbars belong to the CommandBar class. A diamond, line, and star connect CommandBar with CommandItem. That means a CommandBar is composed of CommandItems (or, as you've known all along, a toolbar is made up of commands).

CommandBars is a collection object that keeps track of every toolbar and its commands. You use its Find method (on the ICommandBars interface) to get the ICommandItem interface of a command or toolbar. The Find method has an argument for specifying the command item or command bar you want. You get the CommandBars object, with IDocument's CommandBars property.

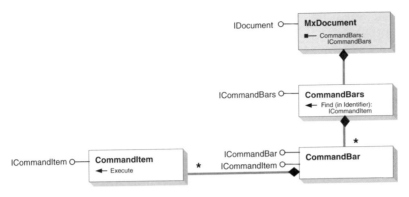

The Find method takes an Identifier number as its required argument. Every command item and command bar—in fact, every COM class—has such a unique identification number. This number is called a GUID (rhymes with squid and stands for "globally unique identifier") or sometimes just a UID.

GUIDs for all the ArcMap commands are listed in a table in the developer help under the help topic *ArcMap: Names and IDs of commands and commandbars.* (You can find the table by doing a search for *ArcMap IDs.*) In the table, buttons are listed by their tooltip name in the Caption column and GUIDs are listed in the GUID column.

Captions   GUIDs

Type	Caption	Name	Command Category	GUID (CLSID / ProgID)
Toolbar	Main Menu	Main Menu	none	{1E739F59-E45F-11D1-9496-080009EE esriCore.MxMenuBar
Menu	File	File_Menu	none	{56599DD3-E464-11D1-9496-080009EF esriCore.MxFileMenu
Command	New	File_New	File	{119591DB-0255-11D2-8D20-080009EF esriCore.MxFileMenuItem
Command	Open	File_Open	File	{119591DB-0255-11D2-8D20-080009EF esriCore.MxFileMenuItem
Command	Save	File_Save	File	{119591DB-0255-11D2-8D20-080009EF esriCore.MxFileMenuItem
Command	Save As	File_SaveAs	File	{119591DB-0255-11D2-8D20-080009EF esriCore.MxFileMenuItem
Command	Add Data	File_AddData	File	{E1F29C6B-4E6B-11D2-AE2C-080009E(esriCore.AddDataCommand

Since you want to execute the Fixed Zoom Out command, you scroll down in the ArcMap IDs table to locate it.

Fixed Zoom Out button   Fixed Zoom Out button's GUID

Toolbar	Tools		Tools_Toolbar	none	{E1F29C75-4E6B-11D2-AE2C-080009EC732A} esriCore.BrowseToolBar
Command	Zoom In		PanZoom_ZoomIn	Pan/Zoom	{AD1891E4-7C79-11D0-8D7C-0080C7A4557D} esriCore.ZoomInTool
Command	Zoom Out		PanZoom_ZoomOut	Pan/Zoom	{F89EBCDF-967C-11D1-873A-0000F8751720} esriCore.ZoomOutTool
Command	Fixed Zoom In		PanZoom_ZoomInFixed	Pan/Zoom	{0830FB33-7EE6-11D0-87EC-080009EC732A} esriCore.ZoomInFixedCommand
Command	Fixed Zoom Out		PanZoom_ZoomOutFixed	Pan/Zoom	{0830FB34-7EE6-11D0-87EC-080009EC732A} esriCore.ZoomOutFixedCommand

To get the command, you could type in the GUID as the Find method's Identifier argument. However, GUIDs are made up of thirty-two numbers and letters, and are a great place to make mistakes. Fortunately, you never have to type a GUID because ESRI has written procedures to get them for you. These procedures are stored in the ArcID code module found in every Normal.mxt project.

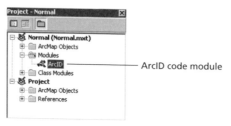

ArcID code module

The ArcID module is a class module filled with property procedures, each of which, given a command's name, returns a different GUID. You may find it more convenient to think of it as an object with a different property for each command.

In the ArcID code module below, you see a property procedure called PanZoom_ ZoomOutFixed and in the procedure you see the Fixed Zoom Out button's GUID.

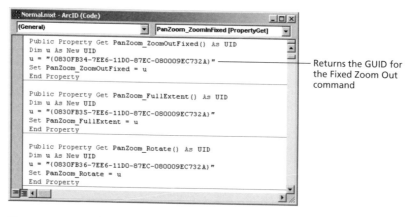

Returns the GUID for the Fixed Zoom Out command

You can write a line of code to get the GUID, using ArcID as an object and the command's name (as it appears in the Name column of the table) as the property. The line of code below returns the Fixed Zoom Out command's GUID.

```
ArcID.PanZoom_ZoomOutFixed
```

You would never open the ArcID module and scroll through its hundreds of procedures to find the one you want. Instead you would look in the ArcMap IDs table, where all the names are listed in its third column.

Fixed Zoom Out button's name

Toolbar	Tools	Tools_Toolbar	none	{E1F29C75-4E6B-11D2-AE2C-080009EC732A} esriCore.BrowseToolBar
Command	Zoom In	PanZoom_ZoomIn	Pan/Zoom	{AD1891E4-7C79-11D0-8D7C-0080C7A4557D} esriCore.ZoomInTool
Command	Zoom Out	PanZoom_ZoomOut	Pan/Zoom	{F89EBCDF-967C-11D1-873A-0000F8751720} esriCore.ZoomOutTool
Command	Fixed Zoom In	PanZoom_ZoomInFixed	Pan/Zoom	{0830FB33-7EE6-11D0-87EC-080009EC732A} esriCore.ZoomInFixedCommand
Command	Fixed Zoom Out	PanZoom_ZoomOutFixed	Pan/Zoom	{0830FB34-7EE6-11D0-87EC-080009EC732A} esriCore.ZoomOutFixedCommand

Since the Find method requires a GUID, you can insert this line of code as Find's Identifier argument, as shown below.

```
Dim pCommandItem As ICommandItem
Set pCommandItem = CommandBars.Find _
    (ArcID.PanZoom_ZoomOutFixed)
```

Once you have the command item, you can run its Execute method.

```
pCommandItem.Execute
```

This last line makes the Fixed Zoom Out button's source code run. The map zooms out, just as if the user had clicked the button on the interface.

You get toolbars in the same way. Suppose you do a lot of editing in a certain map document, and you want the Editor toolbar to appear whenever you open that .mxd file. Just as you do with commands, you use ArcID and a name to get a toolbar's GUID. In the ArcMap IDs table, the Editor toolbar's name is "Editor_EditorToolbar."

Editor toolbar's caption   Editor toolbar's name

| Editor | Editor_EditorToolbar | none | {C671B640-83B9-11D2-850C-0000F875B9C6} esriCore.EditorToolBar |

13
14
15
16
17
18
19
20

In the Find method, you use ArcID and the toolbar's name to return its GUID.

```
Set pCommandItem = CommandBars.Find _
    (ArcID.Editor_EditorToolbar)
```

Once you have gotten the toolbar, working with it is a little different. Toolbar properties and methods are located on the ICommandBar interface, not on ICommandItem, so you do QueryInterface to access them.

```
Dim pCommandBar As ICommandBar
Set pCommandBar = pCommandItem
```

The ICommandBar interface is shown below. Its Dock method has a dockFlags argument that takes values like esriDockFloat, esriDockHide, esriDockLeft, esriDockRight, esriDockShow, esriDockToggle, and esriDockTop. These values dock the toolbar in different positions, hide it, and toggle its visibility.

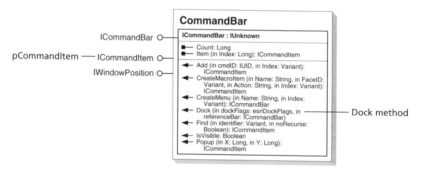

So if you wanted to make the Editor toolbar open in a centered floating position, you would write:

```
pCommandBar.Dock esriDockFloat
```

Exercise 13

You are a historian working with old maps of a hundred or so different cities. Each map has been scanned into an image file. Your project involves georeferencing each city's current data to the scanned maps to see what changes have occurred over time.

To begin to georeference two data sets, you make a data frame, add the data, and turn on the ArcMap Georeferencing toolbar. Since you have to go through this process about a hundred times (in fiction, not in the exercise), you'll code a UIButton to accomplish all three tasks in one click.

You will look up the GUIDs for the Add Data button, the Data Frame menu choice on the Insert menu, and the Georeferencing toolbar. Then you'll write code to run the Execute method on each of the command items and to open the toolbar.

At the end of the exercise, you'll compare new and old data for the city of Manhattan, Kansas. The actual georeferencing will already have been done for you.

1 Start ArcMap and open **ex13a.mxd** in the **C:\ArcObjects\Chapter13** folder.

The map document itself is empty. On the Standard toolbar, a UIButton has been added for you next to the Add Data button. You will write code for its click event.

New button

2 Right-click the new button and click View Source.

You see the button's empty click event, AddMyData_Click.

Starting from ThisDocument, you will first get the CommandBars property (located on the IDocument interface), which returns the ICommandBars interface.

3 Inside the click event, add the following two lines of code to declare and set a CommandBars variable.

```
Dim pCommandBars As ICommandBars
Set pCommandBars = ThisDocument.CommandBars
```

ICommandBars has the Find method to get commands and toolbars. Before writing any more code, you will locate the GUID for the Data Frame menu choice on the Insert menu, shown below.

Data Frame command

4 In the code window, double-click ICommandBars to highlight it.

You are going to open the online help.

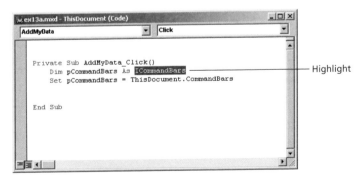

Highlight

5 Press F1 on the keyboard to open the ArcGIS developer help.

The help opens to the ICommandBars Interface topic, but that's not what you want.

6 In the help window, with the Search tab active, check Search titles only and search for **ArcMap IDS**. (You may have to click Display to show the topic.)

7 Scroll down the help page to the beginning of the table. Depending on extensions you have loaded, your list may be different.

Type	Caption	Name	Command Category	GUID (CLSID / ProgID)
Toolbar	Main Menu	Main Menu	none	{1E739F59-E45F-11D1-9496-080009EEBECB} esriCore.MxMenuBar
Menu	File	File_Menu	none	{56599DD3-E464-11D1-9496-080009EEBECB} esriCore.MxFileMenu
Command	New	File_New	File	{119591DB-0255-11D2-8D20-080009EE4E51} esriCore.MxFileMenuItem
Command	Open	File_Open	File	{119591DB-0255-11D2-8D20-080009EE4E51} esriCore.MxFileMenuItem
Command	Save	File_Save	File	{119591DB-0255-11D2-8D20-080009EE4E51} esriCore.MxFileMenuItem
Command	Save As	File_SaveAs	File	{119591DB-0255-11D2-8D20-080009EE4E51} esriCore.MxFileMenuItem
Command	Add Data	File_AddData	File	{E1F29C6B-4E6B-11D2-AE2C-080009EC732A} esriCore.AddDataCommand

In the table, toolbars are blue, menus are gray, and commands are white.

8 Keep scrolling until you come to the Insert menu.

"Data Frame" is the menu choice's caption. (The caption reflects the way the command appears on the user interface.) "PageLayout_NewMap" is its name. You use the name with the ArcID module to get the GUID.

Caption   Name

Menu	Insert	Insert_Menu	none	{119591(esriCore.N
Command	Data Frame	PageLayout_NewMap	Page Layout	{C22579(esriCore.N
Command	Title	PageLayout_InsertTitle	Page Layout	{EB70D0(esriCore.I
Command	Text	PageLayout_InsertText	Page Layout	{EB70D0(esriCore.I
Command	Neatline	PageLayout_Neatline	Page Layout	{F0877F(esriCore.N
Command	Legend	PageLayout_NewLegend	Page Layout	{99D21D esriCore.N
Command	North Arrow	PageLayout_NewNorthArrow	Page Layout	{99D21D esriCore.N
Command	Scale Bar	PageLayout_NewScaleBar	Page Layout	{99D21D esriCore.N
Command	Scale Text	PageLayout_NewScaleText	Page Layout	{99D21D

9 Leave the help window open and bring Visual Basic Editor forward.

10 Declare and set a CommandItem variable using the Find property, ArcID, and the newly found name of the Data Frame menu choice.

```
Dim pCommandItem As ICommandItem
Set pCommandItem = _
    pCommandBars.Find(ArcID.PageLayout_NewMap)
```

Find returns the ICommandItem interface of the specified command.

11 Use ICommandItem's Execute method to run the command's code.

```
pCommandItem.Execute
```

```
ex13a.mxd - ThisDocument (Code)
AddMyData                    Click

Private Sub AddMyData_Click()
    Dim pCommandBars As ICommandBars
    Set pCommandBars = ThisDocument.CommandBars

    Dim pCommandItem As ICommandItem
    Set pCommandItem = _
        pCommandBars.Find(ArcID.PageLayout_NewMap)
    pCommandItem.Execute
End Sub
```

Next you will look up the name of the Add Data button.

12 Bring the help window forward. Scroll down to locate the Standard toolbar (blue) and the Add Data command beneath it.

The command's caption is "Add Data" and its name is "File_AddData."

Toolbar	Standard	Standard_Toolbar	none	{5DEB1D esriCore.S
Command	New	File_New	File	{A33D94 esriCore.F
Command	Open	File_Open	File	{11959{ esriCore.N
Command	Save	File_Save	File	{11959{ esriCore.N
Command	Print	File_Print	File	{11959{ esriCore.N
Command	Cut	Edit_Cut	Edit	{A33D94 esriCore.E
Command	Copy	Edit_Copy	Edit	{A33D94 esriCore.E
Command	Paste	Edit_Paste	Edit	{A33D94 esriCore.E
Command	Delete	Edit_Clear	Edit	{16CD71 esriCore.E
Command	Undo	Edit_Undo	Edit	{FBF8C3F esriCore.N
Command	Redo	Edit_Redo	Edit	{FBF8C3F esriCore.N
Command	Add Data	File_AddData	File	{E1F29C(esriCore.F

13 Leave the help window open and bring Visual Basic Editor forward.

14 Set the pCommandItem variable equal to the Add Data command using the Find property.

```
Set pCommandItem = _
    pCommandBars.Find(ArcID.File_AddData)
```

You don't need to declare a second ICommandItem variable. You can just reassign pCommandItem to point to the Add Data command's ICommandItem interface.

15 Use ICommandItem's Execute method to run the command's code.

```
pCommandItem.Execute
```

Next you will get the GUID for the Georeferencing toolbar.

16 Bring the help window forward. Scroll down to locate the Georeferencing toolbar.

Its caption is "Georeferencing" and its name is "Georeferencing_Toolbar."

Toolbar	Georeferencing	Georeferencing_ToolBar	none	{46B1C. esriCore
Menu	Georeferencing	Georeferencing_Menu	none	{669C6: esriCore
Command	Update Georeferencing	Georeferencing_SaveItem	Georeferencing	{669C6: esriCore
Command	Rectify	Georeferencing_SaveAsItem	Georeferencing	{669C6: esriCore
Command	Fit To Display	Georeferencing_FitToDisplayItem	Georeferencing	{726AD esriCore
Menu	Flip or Rotate	Flip or &Rotate	none	{81751E esriCore
Command	Rotate Right	Georeferencing_RotateRightItem	Georeferencing	{79AE8· esriCore
Command	Rotate Left	Georeferencing_RotateLeftItem	Georeferencing	{79AE8· esriCore

17 Leave the help window open and bring Visual Basic Editor forward.

18 Set the pCommandItem variable equal to the Georeferencing toolbar using the Find property.

```
Set pCommandItem = _
    pCommandBars.Find(ArcID.Georeferencing_Toolbar)
```

As in step 14, you can keep using the same object variable.

Now that you have the toolbar, you want to run its Dock method so that it opens in the desired position when the AddMyData button is clicked. Since the Dock method is on the ICommandBar interface (and the Find method returns ICommandItem), you need to switch interfaces.

19 Declare an ICommandBar variable and set it equal to the ICommandItem variable.

```
Dim pGRCommandBar As ICommandBar
Set pGRCommandBar = pCommandItem
```

20 Use the Dock method to open the toolbar as a floating toolbar.

You will set the DockFlags argument to esriDockFloat, which opens the toolbar in a floating position over ArcMap. You can read about the other esriDockFlag options in the online help.

```
pGRCommandBar.Dock esriDockFloat
```

The code is ready to test.

21 Close Visual Basic Editor and the help window.

22 On the Standard toolbar in ArcMap, click the AddMyData button. In the Add Data dialog box, navigate to the **C:\ArcObjects\Data\Manhattan_KS** folder. Highlight CityOutline2003.lyr, manhattan1890.sid, and Streets.lyr, as shown in the graphic below. Click Add.

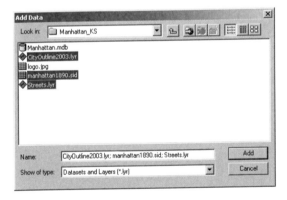

The layers are added to a new data frame. The Georeferencing toolbar floats above the ArcMap application. (Yours may be in a different position than the one in the graphic.)

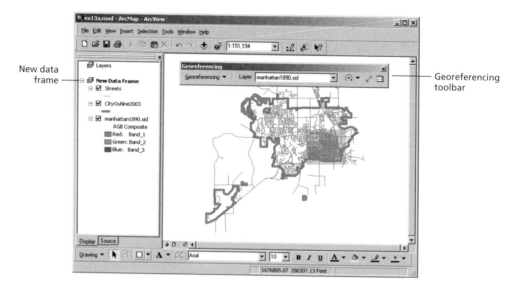

New data frame

Georeferencing toolbar

23 In the table of contents, right-click manhattan1890.sid and click Zoom to Layer.

24 Zoom in on the 2nd Ward on the east-central side of manhattan1890.sid.

1887 Manhattan, Kansas, map, courtesy of the David Rumsey Map Collection, www.davidrumsey.com

Surprisingly, the map compiled in the 1800s matches well with the streets of 2003, but notice that where the Big Blue River once flowed there are now streets. Over the last century, several floods have caused the river channel to migrate eastward, allowing development where the river once flowed.

25 If you want to save your work, click the File menu in ArcMap and click Save As. Navigate to **C:\ArcObjects\Chapter13**. Rename the file **my_ex13.mxd** and click Save. If you are continuing with the next chapter, leave ArcMap open. Otherwise close it.

Adding layers to a map

Adding a geodatabase feature class

Adding a raster data set

You might think the process of adding a layer to a map goes pretty fast, and it probably does if all your data is in one folder. But what if the data is scattered across many computers? What if you connect to ArcSDE® databases and enter password and connection information? What if you get data from Web sites such as Geography Network™? Navigating to data locations, entering passwords, and browsing Web sites takes time. What if everyone in your organization has to do the same?

In this chapter, you will make some layers available from a menu, so that users can just click to add them to a map, and save themselves the time of navigating to the data. Your code will perform four steps. First, it will create a layer from one of the many layer coclasses below; second, it will get a data set from a storage location on a computer; third, it will associate the data set with the layer; and, fourth, it will add the layer to a map.

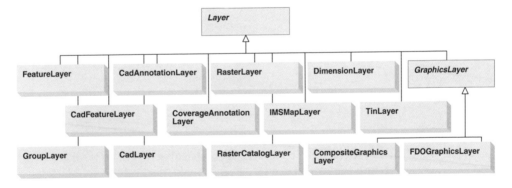

The first step is easy because new layers are created from coclasses in the usual way. The type of layer you create depends on the data source you are going to associate with it. Say the layer's data source will be an ArcInfo grid. A grid is a raster data set, so you make a raster layer.

```
Dim pRLayer As IRasterLayer
Set pRLayer = New RasterLayer
```

Or say your data source is a shapefile. Shapefiles are feature data sets, so you make a feature layer.

```
Dim pFLayer As IFeatureLayer
Set pFLayer = New FeatureLayer
```

The second step is more involved. ArcGIS data sets come in a variety of file formats, but they all have one thing in common—they are stored somewhere on a computer. That storage location is called a workspace. It may be a folder (for shapefiles), a Microsoft Access file (for geodatabase feature classes), a pair of linked folders (for ArcInfo coverages and grids), or even a relational database or a Web site.

To get a data set, no matter how it is stored, you first get its workspace. Workspaces, however, are regular classes, so you can't make or get them yourself. Instead, you get a workspace from a coclass called a workspace factory. So the process is to create a workspace factory, use it to get a workspace, and then get the data set you want from the workspace.

The following diagram illustrates the class relationships. Workspace factories get workspaces. Workspaces are composed of data sets. Dataset is an abstract class with many subclasses. Only one (FeatureClass) is shown here, but there are others for raster data sets, standalone tables, and so on.

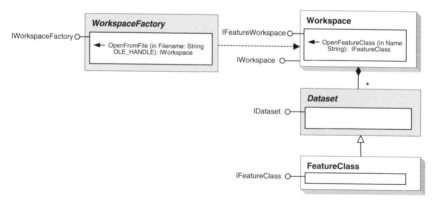

Like the Layer class shown above, WorkspaceFactory is an abstract class with many coclasses, and the workspace factory you make depends on the data set you want to get.

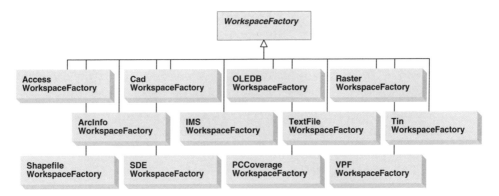

If you compare the workspace factory diagram to the layer diagram, you'll see that the classes don't correspond exactly. For example, you use different workspace factories to get shapefiles (ShapefileWorkspaceFactory) and geodatabase feature classes (AccessWorkspaceFactory), but both are added as feature layers.

When you have got a workspace from a workspace factory, you get a data set from the workspace. Dataset is yet another abstract class with different subclasses for different types of data. For example, feature-based data sets belong to a class called FeatureClass, while raster-based data sets belong to a class called RasterDataset.

After this second step, the third and fourth steps are again pretty easy. It takes one line of code to associate a layer with a data set and one more to add the layer to a map.

In the next two exercises, you'll go through the process of adding layers with two different kinds of data: first, a geodatabase feature class, and then a raster. You'll see that the process is similar, but that some of the classes, interfaces, properties, and methods you use vary.

Adding a geodatabase feature class

When you add a layer to ArcMap, you need to know the format of the data that will go with it, since this affects your code. As long as you know the data type, it doesn't matter whether you create the layer first or get the data first.

Say you have a Streets feature class in a personal geodatabase, and you want to make a layer from it. You might decide to create the layer first.

```
Dim pFLayer As IFeatureLayer
Set pFLayer = New FeatureLayer
```

With that done, you can turn your attention to the data. The process of getting a data set begins with creating a workspace factory. Personal geodatabases are stored in Microsoft Access file format, so in this situation, you want a Microsoft Access workspace factory.

```
Dim pAWFactory As IWorkspaceFactory
Set pAWFactory = New AccessWorkspaceFactory
```

Although you are creating a Microsoft Access workspace factory, the variable is declared to IWorkspaceFactory on the abstract WorkspaceFactory class. (Class inheritance says it's okay to do this.) You want IWorkspaceFactory because it has a method called OpenFromFile that gets workspaces.

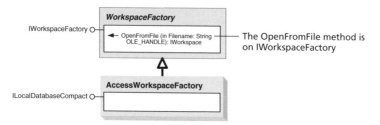

You use the OpenFromFile method to specify the path to the workspace. (In spite of its name, OpenFromFile is used to get any kind of workspace, whether it's a file, a folder, a database, or whatever.) As the preceding diagram shows, OpenFromFile's first argument is a file name as a string. In this case, the string is the path to an .mdb file. So if the Streets feature class was stored in a geodatabase called City.mdb in the D:\Data folder, you would get the workspace with the following code:

```
Dim pFWorkspace As IFeatureWorkspace
Set pFWorkspace = pAWFactory.OpenFromFile _
    ("D:\Data\City.mdb",0)
```

OpenFromFile's second argument is a window handle (or OLE_HANDLE). This is a number assigned by the operating system to identify each open window on your computer. The value 0 is not the actual handle, but a default number that tells VBA to get the ArcMap window handle for you. You can also get the handle with the IApplication interface's hWnd property (Application.hWnd).

Window handles are important when you get workspaces that have connection properties. For example, to get an ArcSDE workspace, you have to set a user name and password. If the connection fails, the OpenFromFile method pops up a Connection dialog box to prompt for the correct information. That dialog box needs to know who its "parent" window is (which application window it belongs to), so that it displays in front of that window and opens and closes with that window. The window handle provides that information.

As a VBA programmer, you don't have to worry about window handles because the only one you need is that of ArcMap, and VBA gets it for you. Programmers working outside VBA, however, may access workspaces from custom application windows that they themselves create. In that case they must get their window's handle to pass to the OpenFromFile method.

If you look again at the diagram above, you'll see that OpenFromFile returns IWorkspace. Then why is the variable declared to IFeatureWorkspace? Here is a case where you can apply the shortcut you learned in chapter 12. You don't need IWorkspace, but you do need IFeatureWorkspace, which has an OpenFeatureClass method to get the data set from the workspace. So you declare the variable to the interface you want and let VBA do QueryInterface for you.

Now that you have the workspace, you can get the data set by running the OpenFeatureClass method. The diagram above shows that this method takes a string as its argument. The string is the name of the feature class, which in this case is Streets.

```
Dim pFClass As IFeatureClass
Set pFClass = pFWorkspace.OpenFeatureClass("Streets")
```

That takes care of the first two steps in the process. You still have to associate the layer you made with the data set and then add the layer to ArcMap.

You associate a feature class with a layer by setting the FeatureClass property on IFeatureLayer. (You have a variable pointing to this interface from when you created the layer.)

The FeatureClass property is different from other properties you have used so far. In the diagram above, you see that the right side of the barbell is open instead of solid.

Properties with two solid barbells, like the ones you're used to, are called Set By Value properties (or byVal for short). Properties with an open barbell on the right are called Set By Reference properties (byRef for short).

The technical differences between these two kinds of properties don't really affect VBA programmers. What you need to know is that the code for setting them is slightly different. To set a byRef property, you use the Set keyword at the start of the line.

```
Set pFLayer.FeatureClass = pFClass
```

BYVAL AND BYREF PROPERTIES

A property that is set by value does not change if the object used to set it is later assigned a new value. For example, say you have a color object called pColor that stores a certain shade of blue. You go on to use that color object to set a symbol's color property with a line of code like "pSymbol.Color = pColor." Now say that you replace pColor's blue value with a shade of red. The symbol remains blue (as long as you don't run the code again to reset the property).

A property that is set by reference changes whenever the object that sets it is assigned a new value. If the Color property were set by reference, then the symbol would become red when you changed the value of pColor.

Technically, when a byVal property is set, VBA makes a copy of the object. The copy is used to set the property. When a byRef property is set, a pointer is used to point to the original object. Thus, if the object's value changes, the property setting changes with it.

Finally, you add the layer to a map. You get a map with the FocusMap property on IMxDocument. The code should be familiar to you from chapters 11 and 12.

```
Dim pMxDoc As IMxDocument
Set pMxDoc = ThisDocument

Dim pMap As IMap
Set pMap = pMxDoc.FocusMap
```

You add the layer with the AddLayer method on IMap.

```
pMap.AddLayer pFLayer
```

Exercise 14a

You are a GIS programmer for the Emergency Management office of Wilson County, North Carolina. During emergencies, analysts in your group make maps for police officers, firefighters, and other public safety officials.

Every emergency is different and so are the maps they require. Analysts might load any number of layers in different data formats from computers across their internal network and the Internet. Currently, to add a layer to a map, an analyst clicks the Add Data button, then navigates through folders and disk connections. During an emergency, this process consumes valuable time.

In this exercise, you will write code so that analysts can add layers directly from a menu. This will be useful for common data sets like streets, parcels, and utility networks. First, you will create the menu itself and a Railroads menu choice for it. Then you will write code to get the data, make the layer, associate the layer with the data, and add the layer to the map.

1 Start ArcMap and open **ex14a.mxd** in the **C:\ArcObjects\Chapter14** folder.

When the map opens, you see the county outline, along with streets and fire stations layers.

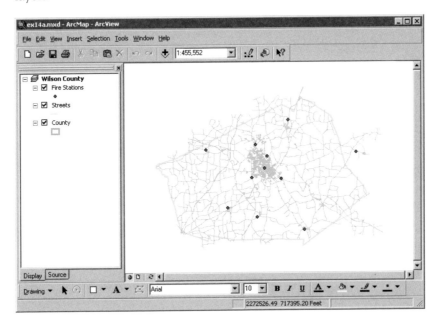

2 In ArcMap, click the Tools menu and click Customize. Click the Commands tab.

3 Make sure that the Save in drop-down list at the bottom of the dialog box is set to ex14a.mxd. In the Categories list, scroll down to the bottom of the list and click [New Menu].

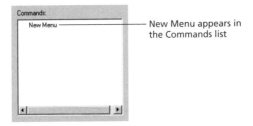

New Menu appears in the Commands list

4 From the Commands list, drag New Menu to the ArcMap main menu. Drop it between the Tools and Window menus.

5 On the ArcMap main menu, right-click New Menu. Highlight the "New Menu" text and replace it with **Add Layers**.

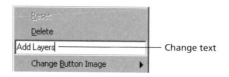

Change text

6 Press Enter.

The new name displays on the menu.

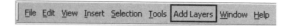

Next, you will add a choice to the menu.

7 In the Customize dialog box, in the Categories list, click UIControls. Make sure that the Save in drop-down list is set to ex14a.mxd.

8 Click New UIControl. In the NewUIControl dialog box, the UIButtonControl option is selected. Click Create.

In the Commands list, you see a new button named Project.UIButtonControl1.

9 In the Commands list, click UIButtonControl1 and change its name to **Project.Railroads**. Press Enter.

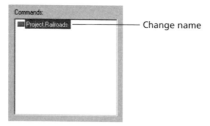

Change name

10 From the Commands list, drag Project.Railroads to the Add Layers menu.

The Railroads choice is added. Next, you will write code for it.

11 Click the Add Layers menu, right-click Railroads, and click View Source.

You see the ThisDocument code module and the empty Railroads click event procedure. Your code here will get a railroads data set, create a new layer, associate the data and the layer, and add the layer to the map.

The data set you want is a feature class within the Wilson County geodatabase (Wilson.mdb). Geodatabases are stored as Microsoft Access files. To get the workspace, then, you make a Microsoft Access workspace factory.

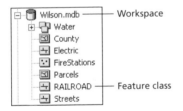

12 In the Railroads click event, declare and set a variable to create a Microsoft Access workspace factory object.

```
Dim pAWFactory As IWorkspaceFactory
Set pAWFactory = New AccessWorkspaceFactory
```

You will get the workspace using the OpenFromFile method on IWorkspaceFactory. This method returns IWorkspace, but the interface you really want is IFeatureWorkspace. You will therefore declare the variable to IFeatureWorkspace and let VBA take care of the QueryInterface.

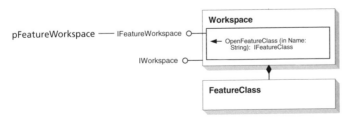

Adding a geodatabase feature class

13 Declare an IFeatureWorkspace variable.

```
Dim pFeatureWorkspace As IFeatureWorkspace
```

The OpenFromFile method requires the full path to the workspace. You could type this in, but it's easy to make mistakes. Instead, you'll open ArcCatalog and copy the path from its Location toolbar.

14 Start ArcCatalog.

15 In ArcCatalog, click the View menu, point to Toolbars, and click Location if it is not already checked.

16 In the ArcCatalog tree, navigate to **C:\ArcObjects\Data\Wilson_NC**. (If you installed the data for this book to a different path, use that path instead.) In the catalog tree, click the Wilson geodatabase.

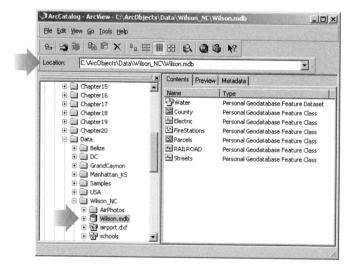

The path to the workspace appears in the Location toolbar's drop-down list.

17 Highlight the path as shown. Right-click in the blue highlighted area and click Copy.

18 Bring Visual Basic Editor forward.

19 Set the workspace variable with the OpenFromFile method. For the first argument, paste the path you copied from ArcCatalog and add quote marks around it. Use 0 for the second argument.

```
Set pFeatureWorkspace = pAWFactory.OpenFromFile _
("C:\ArcObjects\Data\Wilson_NC\Wilson.mdb", 0)
```

Now that you have the workspace, you can get the railroads feature class. To do this, you run the OpenFeatureClass method, which is on IFeatureWorkspace. The argument for OpenFeatureClass is the name of the feature class in quotes. The name must be an exact match, but is not case-sensitive.

20 Declare and set a variable to get the railroad feature class.

```
Dim pFClass As IFeatureClass
Set pFClass = pFeatureWorkspace.OpenFeatureClass _
("Railroad")
```

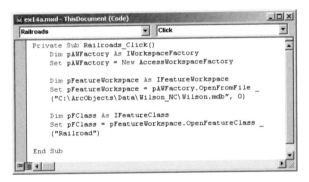

In the introduction to this exercise, you learned that it doesn't matter whether you make the layer first and then get the data, or do it the other way around. In this case, you got the data first. Now it is time to make the layer.

21 Declare and set a variable to create a feature layer.

```
Dim pFLayer As IFeatureLayer
Set pFLayer = New FeatureLayer
```

To associate the layer with the data, you use IFeatureLayer's FeatureClass property. This is a byRef property (open barbell), so it requires the Set keyword.

22 Set the FeatureLayer's FeatureClass property equal to the railroad feature class.

```
Set pFLayer.FeatureClass = pFClass
```

Before adding the layer to the map, you will give it a name. The Name property is found on the abstract Layer class's ILayer interface, shown below. The pFLayer variable points to FeatureLayer's IFeatureLayer interface. Since Layer is FeatureLayer's superclass you can use QueryInterface to switch from IFeatureLayer to ILayer.

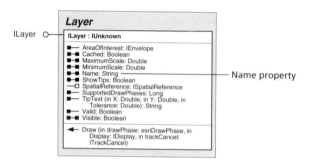

23 Declare an ILayer variable, set it with a QI, and then set the layer's name.

```
Dim pLayer As ILayer
Set pLayer = pFLayer
pLayer.Name = "Railroads"
```

Now you can add the layer to a map.

24 Declare and set an IMxDocument variable.

```
Dim pMxDoc As IMxDocument
Set pMxDoc = ThisDocument
```

25 Get the active map and add the layer to it.

```
Dim pMap As IMap
Set pMap = pMxDoc.FocusMap

pMap.AddLayer pFLayer
```

To draw the layer, you have to refresh the map display. Back in chapter 11, you did this by getting the active view and refreshing it with the following code:

```
'Dim pActiveView as IActiveView
'Set pActiveView = pMxDoc.ActiveView

'pActiveView.Refresh
```

Now you have an opportunity to slip in another shortcut. You can cut these three lines down to one with a technique called chaining. Chaining allows you to extend the object.property syntax with forms like object.property.property and object.property.method.

```
'pMxDoc.ActiveView.Refresh
```

When VBA runs a chained line of code, it evaluates the first object.property part of the statement to see which interface is returned. The next property or method in the chain is then issued to that interface. In the line above, pMxDoc.ActiveView returns IActiveView, which is then issued the Refresh method. As long as each new property or method is appropriate to the interface returned by the previous link, you can take chaining as far as you want.

26 Refresh the map display area and update the table of contents.

```
pMxDoc.ActiveView.Refresh
pMxDoc.UpdateContents
```

13
14
15
16
17
18
19
20

The Railroads UIButton is ready to test.

27 Close Visual Basic Editor.

28 In ArcMap, click the Add Layers menu and click Railroads.

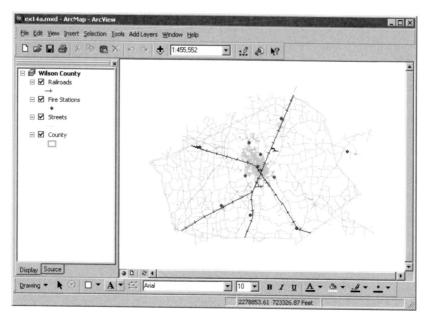

The Railroads layer is added to the map and draws with appropriate symbology.

When the name of a feature class matches the name of an ArcGIS symbol, that symbol is automatically used to draw the layer. As the following graphic shows, the Symbol Selector contains a Railroad symbol. (Note that the symbol name has to match the name of the feature class—not the name of the layer—and the match does not have to be case-sensitive.)

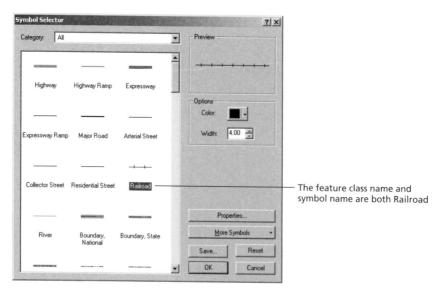

The feature class name and symbol name are both Railroad

You have added the first of many possible menu choices to the Add Layers menu. You will add another one in the next exercise, but this book can't take you through every data format supported by ArcGIS. To help you explore them on your own, you will find a text file on the CD that comes with this book at C:\ArcObjects\Data \Samples\AddDataSubs.txt. The file contains three sample subroutines to create layers from a CAD .dxf file, a shapefile, and an ArcInfo coverage feature class. The corresponding data sets are also included on the CD.

As an experiment, you could create three more menu choices on the Add Layers menu. Then, for each choice, you could call a different procedure from the samples in the text file.

29 If you want to save your work, click the File menu in ArcMap and click Save As. Navigate to **C:\ArcObjects\Chapter14**. Rename the file **my_ex14a.mxd** and click Save. If you are continuing with the next exercise, leave ArcMap open. Otherwise close it.

Adding a raster data set

Working with raster layers is similar to working with feature layers. In both cases, you locate a workspace that contains a data set, get the data set by name, associate the data set with a new layer, and add the layer to a map.

In the previous exercise, you worked with a Microsoft Access workspace, which is an .mdb file. In this exercise, you will work with a raster workspace, which is a folder. In the following graphic, the AirPhotos folder contains several images, each of which is a raster data set. (The image files are in MrSID® format, which has the .sid file extension.)

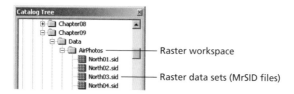

In the previous exercise, you learned that to create or get a workspace you need a workspace factory. For raster data, you begin with a RasterWorkspaceFactory. The diagram below shows the relevant class relationships.

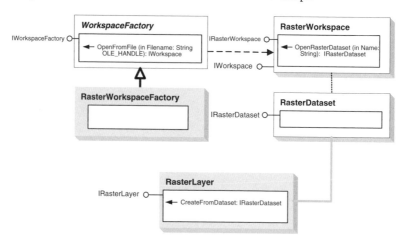

When you make a raster workspace factory, you declare the variable to IWorkspaceFactory, just as you did in the last exercise, because IWorkspaceFactory has the OpenFromFile method.

After getting the workspace, you get the raster data set. When you got the feature class in the last exercise, you switched from IWorkspace to IFeatureWorkspace, which had the OpenFeatureClass method you needed. In this situation, you switch instead to IRasterWorkspace, which has the OpenRasterDataset method. Same idea, different interface and method.

The OpenRasterDataset method takes the name of the raster as its argument.

```
Dim pRDataset As IRasterDataset
Set pRDataset = pRWorkspace.OpenRasterDataset("air.sid")
```

Once you have the data set, you create a raster layer. (Or, if you prefer, do it the other way around.)

```
Dim pRLayer As IRasterLayer
Set pRLayer = New RasterLayer
```

To associate the layer with the data set, you run the CreateFromDataset method on the IRasterLayer interface. Again, it's slightly different from the way things work with feature classes.

```
pRLayer.CreateFromDataset pRDataset
```

You add the raster layer to a map in the usual way.

```
Dim pMxDoc As IMxDocument
Set pMxDoc = ThisDocument

pMxDoc.AddLayer pRLayer
```

Exercise 14b

In emergencies like floods and fires, the area being burned or flooded constantly changes. Air photos help analysts predict what might happen next because they show terrain and familiar reference points.

In this exercise, you will make a submenu of air photos for the Add Layers menu. Then you will code a menu choice to add an air photo to the map.

1 Start ArcMap and open **ex14b.mxd** in the **C:\ArcObjects\Chapter14** folder.

When the map opens, you see the airport, schools, fire stations, and streets layers, plus an orange graphic representing the location of a train crash and a plume of smoke coming from a burning tanker car.

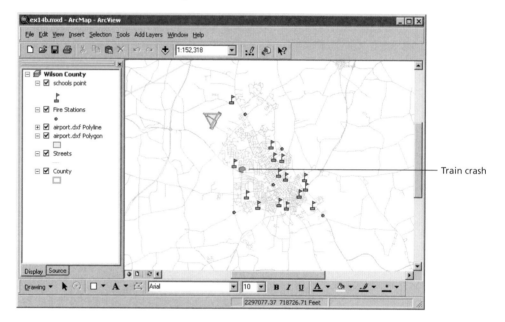

2 Click the Tools menu and click Customize. Click the Commands tab. Make sure that the Save in drop-down list is set to ex14b.mxd.

3 Scroll to the bottom of the Categories list and click New Menu.

You see a New Menu listed in the Commands list.

4 From the Commands list, drag the New Menu to the top of the Add Layers menu. Drop it above Railroads, as shown.

5 Right-click New Menu and replace the text with **Air Photos**. Press Enter.

The new name displays on the menu.

Next you will add a choice (a UIButton) to the Air Photos submenu.

6 In the Customize dialog box, in the Categories list, click UIControls. Make sure that the Save in drop-down list is set to ex14b.mxd.

7 Click New UIControl. In the NewUIControl dialog box, the UIButtonControl option is selected. Click Create.

In the Commands list, you see a new button named Project.UIButtonControl1, as well as the Project.Railroads button.

8 In the Commands list, click on Project.UIButtonControl1 and rename it by typing **Project.WilsonWest**. Press Enter.

9 From the Commands list, drag Project.WilsonWest to the Air Photos submenu.

10 Right-click WilsonWest and click View Source.

You see the ThisDocument code module and the empty WilsonWest click event procedure. You will write code to get an image file from a raster workspace and add it as a layer to ArcMap.

This time, your starting point is the RasterWorkspaceFactory coclass.

11 In the click event, add the following two lines of code to create a raster workspace factory.

```
Dim pRWFactory As IWorkspaceFactory
Set pRWFactory = New RasterWorkspaceFactory
```

12 Declare and set a variable to get the workspace using the OpenFromFile method. (If you installed the data for this book to a different path, use that path instead.)

```
Dim pRasterWorkspace As IRasterWorkspace
Set pRasterWorkspace = pRWFactory.OpenFromFile _
("C:\ArcObjects\Data\Wilson_NC\AirPhotos", 0)
```

OpenFromFile returns IWorkspace, but here again, you have declared the variable to the interface you really want. In this case, it's IRasterWorkspace, which has the OpenRasterDataset method.

13 Get the raster data set using the OpenRasterDataset method.

```
Dim pRDataset As IRasterDataset
Set pRDataset = pRasterWorkspace.OpenRasterDataset _
("wilsonwest.sid")
```

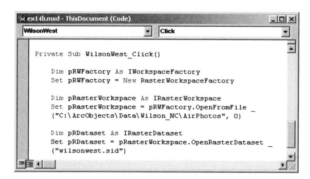

Now that you've got a data set, you can make a new raster layer and then associate the two.

14 Create a new raster layer.

```
Dim pRLayer As IRasterLayer
Set pRLayer = New RasterLayer
```

To associate the layer with the data set, you run the CreateFromDataset method. Running this method on a raster layer is equivalent to setting the FeatureClass property on a feature layer.

15 Add a line to associate the raster data set with the raster layer.

```
pRLayer.CreateFromDataset pRDataset
```

16 Add another line to set the layer's name.

```
pRLayer.Name = "Wilson West"
```

The Name property is on ILayer. The pRLayer variable points to IRasterLayer. You would normally have to QI from IRasterLayer to ILayer and then set the Name property. In this case, a shortcut called interface inheritance is available. You will learn about it soon in chapter 15.

Now you will get the active map from the map document and add the layer to it.

17 Declare and set an IMxDocument variable.

```
Dim pMxDoc As IMxDocument
Set pMxDoc = ThisDocument
```

18 Get the active map and add the layer to it.

```
Dim pMap As IMap
Set pMap = pMxDoc.FocusMap

pMap.AddLayer pRLayer
```

You can't see through the image, so you need to make sure it goes to the bottom of the table of contents.

19 Use the map's MoveLayer method to move the raster layer to the bottom of the table of contents.

```
pMap.MoveLayer pRLayer, pMap.LayerCount - 1
```

MoveLayer has two arguments, the name of the layer you want to move, and the index position you want to move it to. The first index position, at the top of the table of contents, is 0.

Since a user can freely add or remove layers, you have no way of knowing how many layers will be on the map. So how can you move a layer to the bottom? Map has a LayerCount property that returns the number of layers on the map. LayerCount minus 1 returns the position of the bottom layer.

20 Add two lines of code to refresh the map's active view area and the table of contents.

```
pMxDoc.ActiveView.Refresh
pMxDoc.UpdateContents
```

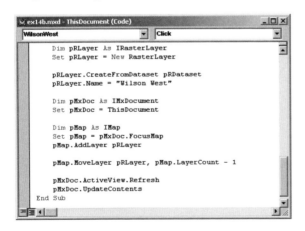

chapter

13
14
15
16
17
18
19
20

The menu choice is ready to test.

21 Close Visual Basic Editor.

There has been a train crash on the northwest side of the city. You will use the new menu choice to create a map of the crash site.

22 Click the View menu, point to Bookmarks, and click Train wreck overview.

The map zooms to the train wreck area. Two polygons show the crash site (pale orange) and a plume of smoke (dark orange) coming from a burning tanker car.

23 Click the Add Layers menu and click Railroads to add the layer to the map.

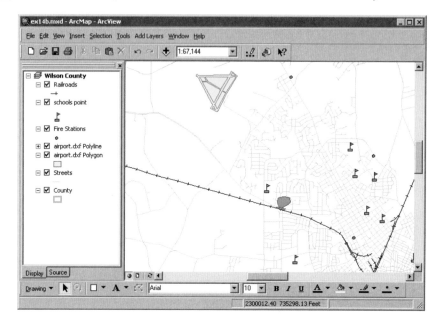

The airport is probably far enough away, but a neighborhood and a school only blocks away will probably have to be evacuated with the help of nearby fire stations.

24 Click the View menu, point to Bookmarks, and click Train wreck detail.

The map zooms in on the area right around the crash.

25 Click the Add Layers menu, point to Air Photos, and click Wilson West.

It looks like the plume is already within a residential area and is nearing the school to the northwest. The area needs to be evacuated until the plume clears.

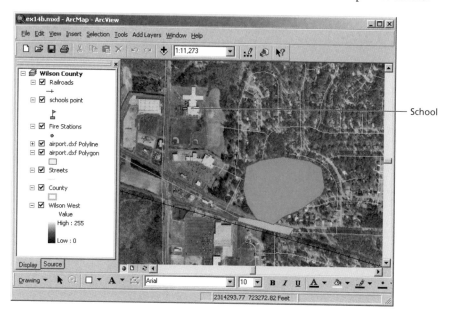

26 If you want to save your work, click the File menu in ArcMap and click Save As. Navigate to **C:\ArcObjects\Chapter14**. Rename the file **my_ex14b.mxd** and click Save. If you are continuing with the next chapter, leave ArcMap open. Otherwise close it.

Setting layer symbology

Setting layer color

Setting layer symbols

Creating a class breaks renderer

Layers on a map have instructions that define the symbols and colors used to draw features. In the map of Mexico below, the Roads layer draws as a wide red line. The Cities layer draws capitals as green stars and other cities as green circles. The States layer represents population density in shades of yellow to brown.

Users tell a layer how to draw by working with layer legends. Programmers call the legend a *renderer* and control it with code. In this chapter, you'll learn how to symbolize layers by making renderers and setting their properties.

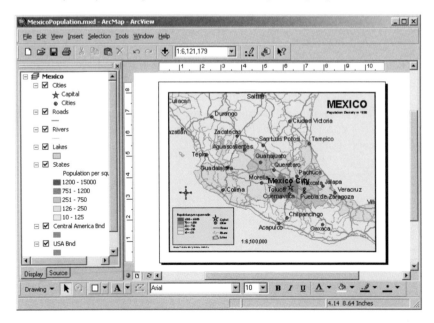

Back in chapter 12, you learned that VBA can do QueryInterface for you, and you've used this shortcut in a couple of places. In this chapter, you will use another coding shortcut called interface inheritance.

You already know about class inheritance, which means that a class has all the interfaces of its superclasses. For example, the FeatureLayer class has all the interfaces of its superclass Layer. With interface inheritance, one interface inherits the properties and methods of another interface. In looking at different diagrams in this book, you may have noticed that each interface's name appears at the top of its white box, followed by a colon and the name of another interface. This notation means that the first interface inherits from the second, as shown below.

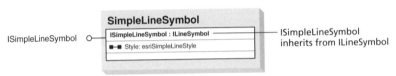

The properties on ILineSymbol can be used as if they were on ISimpleLineSymbol.

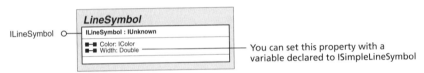

If you have a variable pointing to ISimpleLineSymbol, you can use it to set any ILineSymbol property.

```
Dim pSimpleLineSymbol As ISimpleLineSymbol
Set pSimpleLineSymbol = New SimpleLineSymbol
pSimpleLineSymbol.Width = 3
```

Interface inheritance can only be used with a few interfaces. Most interfaces, like ILineSymbol, above, inherit from an interface called IUnknown, which sits at the top of the interface hierarchy (all classes have it). Although it has no practical use for most VBA programmers, IUnknown provides every class with QueryInterface capability. You can read more about IUnknown in the developer help and at the msdn.microsoft.com Web site.

Setting layer color

When a user adds a feature layer to a map, ArcMap assigns a simple renderer that draws the layer in a single randomly selected color. When a programmer adds a layer to a map, they can use this same default renderer or write code to give the layer different drawing instructions.

The diagram below shows that every feature layer has a renderer, every renderer is composed of symbols, and every symbol has a color. Renderer, Symbol, and Color are all abstract classes, each of which has many subclasses.

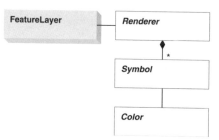

When you create a renderer, it doesn't have any symbols. When you create a symbol, it doesn't have a color. So the best way to get started is to make colors. Then you can make symbols and assign the colors to them. Finally, you make a renderer and add the symbols to it.

In chapter 11, you learned about the Color class and its five color model subclasses. You created an RgbColor object with the code below.

```
Dim pSalmon As IRgbColor
Set pSalmon = New RgbColor
```

Each color object has properties you set to get your desired color. With the RGB color model, you set different amounts of red, green, and blue.

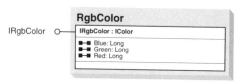

To make a salmon color, for example, you would set the properties as follows:

```
pSalmon.Red = 255
pSalmon.Green = 160
pSalmon.Blue = 122
```

To find out more about color models, names, and values, navigate to your C:\ArcObjects\Data\Samples folder and open the file colornames.txt.

Colors get assigned to symbols. The diagram below shows the Symbol abstract class and three of its subclasses: MarkerSymbol, LineSymbol, and FillSymbol. These are themselves abstract classes with their own subclasses. LineSymbol subclasses include SimpleLineSymbol, MultiLayerLineSymbol, and PictureLineSymbol.

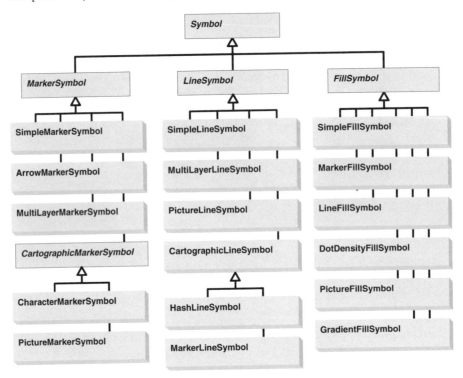

You can create a SimpleLineSymbol with the code below.

```
Dim pLine As ISimpleLineSymbol
Set pLine = New SimpleLineSymbol
```

Some of the other line symbol coclasses do more than draw a simple line. For example, PictureLineSymbol and MarkerLineSymbol let you choose an image file or marker symbol that repeats along the length of the line.

To assign a color to a symbol, you set the Color property on the MarkerSymbol, LineSymbol, or FillSymbol abstract classes. As shown below, LineSymbol has the Color property on its ILineSymbol interface.

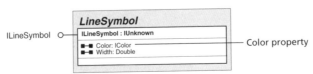

The familiar way to get to this property is to do QueryInterface.

```
Dim pLineSym As ILineSymbol
Set pLineSym = pLine
```

Now you can set the line symbol's color equal to the salmon color you made. You might also set the Width property to draw the symbol in a thick line.

```
pLineSym.Color = pSalmon
pLineSym.Width = 3
```

Alternatively, since ISimpleLineSymbol inherits the ILineSymbol interface, you could accomplish the same thing with less code. pLine points to the ISimpleLineSymbol interface, but can be used to set ILineSymbol's Color and Width.

```
pLine.Color = pSalmon
pLine.Width = 3
```

Having made a color and assigned it to a symbol, you must still assign the symbol to a renderer.

The diagram below shows that FeatureRenderer is connected to FeatureLayer with one line and no other symbols. That means that each FeatureLayer has a FeatureRenderer.

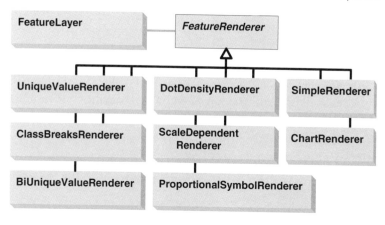

The FeatureRenderer abstract class has eight subclasses, representing different legend types. A SimpleRenderer draws all features in one symbol and one color.

```
Dim pRender As ISimpleRenderer
Set pRender = New SimpleRenderer
```

After creating a renderer, you set its Symbol and Label properties. The Symbol property is a byRef property (open barbell), so you set it with the Set keyword.

The next lines set the renderer's Symbol property to the salmon line symbol, and its Label property to Salmon Streams.

```
Set pRenderer.Symbol = pLineSym
pRenderer.Label = "Salmon Streams"
```

The Label property sets the text that goes with the symbol in the ArcMap table of contents.

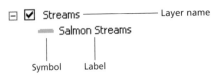

The renderer has been given a symbol, a color, and a label, but it must still be associated with a feature layer. To make this association, you set the Renderer property on the FeatureLayer class's IGeoFeatureLayer interface.

If you haven't already created the feature layer, you make it with the code below. (If the layer already exists, and you have a variable pointing to IFeatureLayer, you would do QueryInterface to IGeoFeatureLayer.)

```
Dim pGFLayer As IGeoFeatureLayer
Set pGFLayer = New FeatureLayer
```

The Renderer property is also a byRef property that requires the Set keyword.

```
Set pGFLayer.Renderer = pRenderer
```

Before you can see the results of your work, you have to refresh the map display and update the table of contents. The code to do this is familiar to you from the last chapter.

Exercise 15a

The Add Layers menu has helped the County's emergency analysts make basemaps much more quickly than before. They have noted one drawback, however. While some layers, like the Railroads layer, are automatically symbolized (because the layer's feature class name matches the name of an ArcGIS symbol), other layers are not. For example, the Water Lines layer draws with default symbology in a random color.

In this exercise, you will modify the code for the Water Lines choice on the Add Layer menu. You will create a renderer, a line symbol, and a color to make the water lines draw in blue.

1 Start ArcMap and open **ex15a.mxd** in the **C:\ArcObjects\Chapter15** folder.

When the map document opens, you see the county outline and its streets.

2 Click the Add Layers menu and click Water Lines.

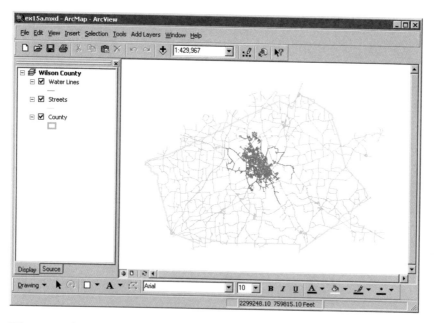

The water lines display in a single randomly selected color. (Your color may be different.)

3 Click the Tools menu and click Customize.

4 With the Customize dialog box open, click the Add Layers menu, right-click Water Lines, and click View Source.

The WaterLines click event already has code to add a water lines layer to the map. It's similar to the code you wrote in chapter 14 to make a layer from a geodatabase feature class and add it to a map. You will add code to set the layer's color.

5 Find the following commented line near the end of the click event procedure.

```
'Add color code here.
```

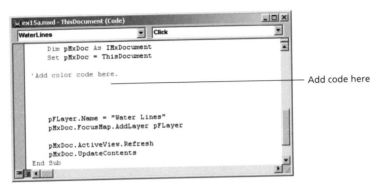

Add code here

6 After the comment, add the following lines to create an RgbColor object.

```
Dim pColor As IRgbColor
Set pColor = New RgbColor
```

7 Set the color's properties to make steel blue.

```
pColor.Red = 70
pColor.Green = 130
pColor.Blue = 180
```

8 Now create a simple line symbol.

```
Dim pLineSym As ISimpleLineSymbol
Set pLineSym = New SimpleLineSymbol
```

To associate the symbol with the color, you need to set the Color property on ILineSymbol. The diagram below shows that ISimpleLineSymbol inherits the ILineSymbol interface, so you can use the interface inheritance shortcut.

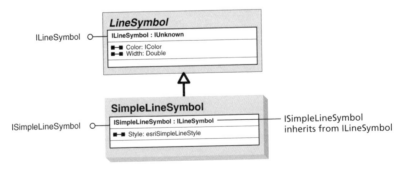

9 Set the simple line symbol's Color property.

```
pLineSym.Color = pColor
```

10 Now create a simple renderer.

```
Dim pRenderer As ISimpleRenderer
Set pRenderer = New SimpleRenderer
```

To associate the renderer with the symbol, you set the simple renderer's Symbol property, which is a byRef property.

11 Set the simple renderer's Symbol property.

```
Set pRenderer.Symbol = pLineSym
```

Now that you have created a renderer, you can assign it to a layer. You do this by setting the layer's Renderer property on the IGeoFeatureLayer interface. As you began this exercise and opened the code window, you saw that much of the click event had already been written. If you look up in the code, you'll see a variable called pFLayer that refers to the WaterLines layer. However, pFLayer points to IFeatureLayer. You need to switch interfaces to IGeoFeatureLayer.

12 Declare an IGeoFeatureLayer variable and switch interfaces.

```
Dim pGFLayer As IGeoFeatureLayer
Set pGFLayer = pFLayer
```

The Renderer property that you will set next is also a byRef property.

13 Assign the new renderer to the Water Lines layer.

```
Set pGFLayer.Renderer = pRenderer
```

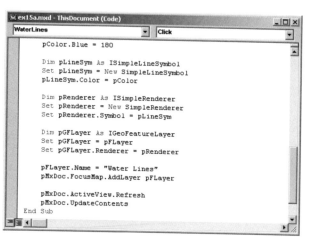

The next four lines of code were already there. They set the layer name, add the layer to the map (using the chaining technique you learned in chapter 14), refresh the map display, and refresh the table of contents, making the new symbology appear.

14 Close Visual Basic Editor.

You will remove the water lines layer that you added in step 2, then add it again to see it draw with the new symbol and color.

15 In the ArcMap table of contents, right-click the Water Lines layer and click Remove.

16 Click Add Layers and click Water Lines.

The layer is symbolized in steel blue in the legend and on the display.

17 For a closer look, right-click Water Lines in the table of contents and click Zoom to Layer.

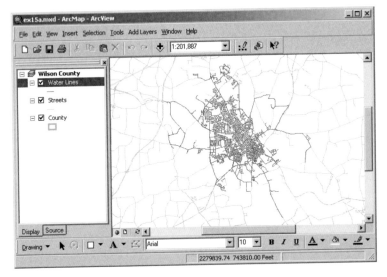

18 If you want to save your work, click the File menu in ArcMap and click Save As. Navigate to **C:\ArcObjects\Chapter15**. Rename the file **my_ex15a.mxd** and click Save. If you are continuing with the next exercise, leave ArcMap open. Otherwise close it.

Setting layer symbols

When you make a feature layer's renderer, you can create your own symbology for it (as you did in the last exercise), but you can also use symbols and colors created by others. Getting symbols that already exist—and that look good—can save you a lot of programming time.

ArcGIS symbols are stored in the Style Manager as styles, style gallery classes, and style gallery items.

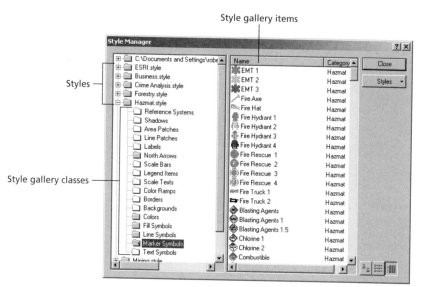

chapter

13
14
15
16
17
18
19
20

The ESRI style contains symbols for all-purpose cartography. The other styles contain symbols unique to an industry or discipline. You can also make your own styles or add styles created by others.

Styles contain style gallery classes, which are groups of similar symbols or map elements. Marker symbols are a style gallery class, and so are line symbols, colors, and north arrows. (Here, a "class" just means a collection of similar things, not an ArcObjects class.)

Style gallery classes contain style gallery items, which are individual symbols or elements. In the right-hand window of the above graphic, you see the items in the Marker Symbols class of the Hazmat (Hazardous Materials) style.

The next graphic shows the items in the Colors style gallery class, again from the Hazmat style. Within a style gallery class, items can belong to different categories, like the ones here.

Category

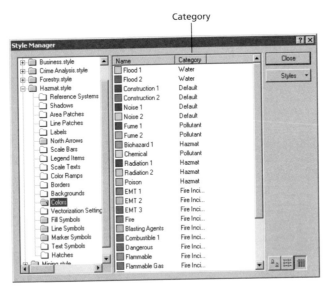

If you have made colors you want to keep for future maps, you can create a new style and save the colors in its Colors style gallery class.

When you want to use a symbol from an existing style, you follow these three steps:

First, you get the style gallery. This is the object that contains all the styles, like ESRI, Business, and Crime Analysis.

Second, you get a list, or enumeration, of style gallery items. The enumeration is called an "Enum" for short.

Third, you get the specific style gallery item you want.

Although a StyleGallery is composed of many StyleGalleryClasses, you don't have to get a style gallery class along the way. As the next diagram shows, you can go straight from the gallery to the Enum to the item. To get the Enum, you use the IStyleGallery interface's Items property. This property has arguments that specify both the style and the style gallery class you want.

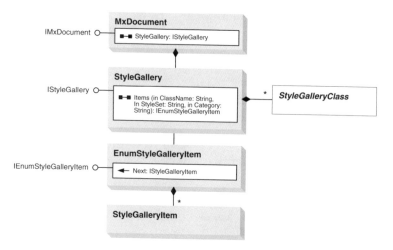

In the following code, the IMxDocument interface's StyleGallery property is used to get the style gallery:

```
Dim pMxDoc As IMxDocument
Set pMxDoc = ThisDocument

Dim pStyleGallery As IStyleGallery
Set pStyleGallery = pMxDoc.StyleGallery
```

The Items property on IStyleGallery returns an Enum of symbols. The Items property has three arguments for specifying the style gallery class, the style, and the symbol category. The code below returns an Enum of line symbols in the ESRI style that belong to the Dashed category.

```
Dim pEnumStyleGallery As IEnumStyleGalleryItem
Set pEnumStyleGallery = pStyleGallery.Items _
        ("Line Symbols", "ESRI.style", "Dashed")
```

The next graphic shows the Enum returned by the code.

···	Contour, Topogra...	Dashed
···	Contour, Bathymet...	Dashed
—	Dashed 6:1	Dashed
--	Dashed 4:1	Dashed
···	Dashed 2:1	Dashed
—	Dashed 6:6	Dashed
- ·	Dashed 4:4	Dashed
···	Dashed 2:2	Dashed
— ·	Dashed 1 Long 1 ...	Dashed
— ·	Dashed 1 Long 2 ...	Dashed

— pEnumStyleGallery

To make your way through the Enum, you use the Next and Reset methods on the EnumStyleGalleryItem coclass.

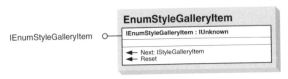

When you first get the Enum, a pointer is pointing to the top of the list (before the first symbol). You move the pointer to get to each symbol.

Pointing to the top of the list ➡

---	Contour, Topogra...	Dashed
---	Contour, Bathymet...	Dashed
---	Dashed 6:1	Dashed
--	Dashed 4:1	Dashed
---	Dashed 2:1	Dashed

The Enum's Next method moves the pointer down one symbol in the list, returning that symbol's IStyleGalleryItem interface.

```
Dim pStyleItem As IStyleGalleryItem
Set pStyleItem = pEnumStyleGallery.Next
```

pStyleItem ➡

---	Contour, Topogra...	Dashed
---	Contour, Bathymet...	Dashed
---	Dashed 6:1	Dashed
--	Dashed 4:1	Dashed
---	Dashed 2:1	Dashed

When you run the Next method on the last symbol in the Enum, the pointer drops off the list and points at a value called Nothing.

---	Contour, Topogra...	Dashed
---	Contour, Bathymet...	Dashed
---	Dashed 6:1	Dashed
--	Dashed 4:1	Dashed
---	Dashed 2:1	Dashed
—	Dashed 6:6	Dashed
-	Dashed 4:4	Dashed
---	Dashed 2:2	Dashed

pStyleItem ➡ Pointing to Nothing at the bottom

You can move the pointer back to the top of the list by running the Reset method.

```
pEnumStyleGallery.Reset
```

To move through the Enum, you put the Next method inside a Do Until loop. The loop below stops when the pointer is at the bottom of the Enum, pointing at Nothing.

```
Do Until pStyleItem Is Nothing
    'Do something to each symbol
    Set pStyleItem = pEnumStyleGallery.Next
Loop
```

How do you know when you are pointing at the symbol you want to get? The IStyleGalleryItem interface has a Name property, which gives you a way to test each item in the Enum. (An item's name is the name that appears in the Style Manager.)

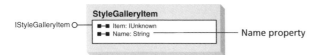

Say you want the Dashed 4:4 line symbol. Inside the loop, you could use an If Then statement to test each item for that name.

```
If pStyleItem.Name = "Dashed 4:4" Then
```

Once you find the symbol, you get it (that is, you get one of its interfaces) using IStyleGalleryItem's Item property, shown in the diagram above. Most likely, you have plans for this symbol, like adding it to a renderer, that require the ILineSymbol interface. The Item property, however, returns IUnknown.

When you learned about interface inheritance at the beginning of this chapter, you learned that every class has IUnknown. The Item property takes advantage of this fact. It is able to return such a wide variety of symbols—including line symbols, marker symbols, colors, and color ramps—because it returns their IUnknown interface.

To get an interface to the Dashed 4:4 symbol, you could declare a variable to IUnknown, set it with the Item property, declare an ILineSymbol variable, and set it by doing QueryInterface from IUnknown to ILineSymbol.

Or, more conveniently, you could let VBA do the QueryInterface for you with the code below. You just declare the variable to ILineSymbol and set it with the Item property.

```
    Dim pLineSym As ILineSymbol
    Set pLineSym = pStyleItem.Item
End If
```

After getting the symbol, you could assign it to a layer's renderer as you did in the previous exercise.

Exercise 15b

The county's emergency analysts have a layer of fire hydrants. Right now, fire hydrants are drawn with a simple point symbol in a random color. In this exercise, you will write code that gets Fire Hydrant 4 from the Hazmat style and assigns it to the layer's renderer. Fire Hydrant 4 looks like a red fire hydrant.

1 Start ArcMap and open **ex15b.mxd** in the **C:\ArcObjects\Chapter15** folder.

When ArcMap opens, you see the county's streets and water lines.

2 Click the Add Layers menu and click Fire Hydrants.

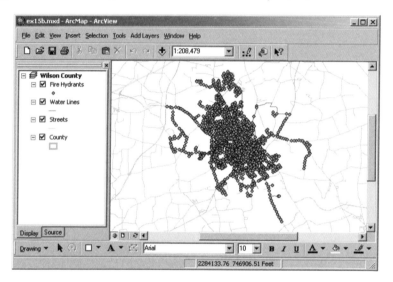

Hydrants display in a random color. The analysts who use the application want the hydrants to look like hydrants. You'll look at the fire hydrant symbols in the Style Manager.

3 Click Tools, point to Styles, and click Style Manager.

The Styles button contains a drop-down list of available styles. A check mark appears to the left of a currently loaded style. Each style can contain hundreds of symbols, so you only load the ones you need.

4 Click the Styles button. In the list of available styles, click Hazmat unless it is already checked. (Depending on your settings, your styles may not match the graphic exactly.)

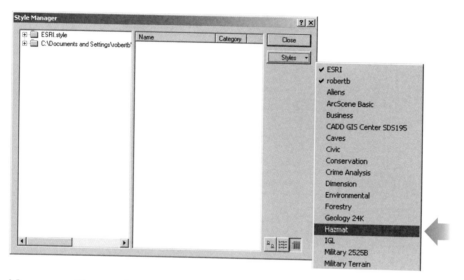

Next you will open the Hazmat style and look at its symbols.

5 In the Style Manager, click the plus sign next to Hazmat.Style to expand its contents.

You see the Hazmat style's style gallery classes represented by folders. A white folder means that a class is empty.

6 Click the Marker Symbols style gallery class folder to open it. Then click the Details button to show the categories.

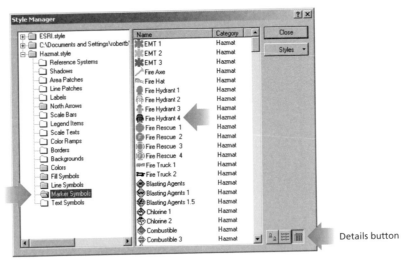

Details button

The Hazmat marker symbols, including several fire hydrant symbols, are displayed by name and category. To get the Enum, you will use the Hazmat style, the Marker Symbols class, and the Hazmat category.

7 Close the Style Manager.

8 Click the Tools menu and click Customize.

9 Click the Add Layers menu, right-click Fire Hydrants, and click View Source.

You see the FireHydrants click event, which adds the Fire Hydrants layer to the map.

10 Scroll down in the code to find the following commented line.

```
'Add hydrant symbol code here.
```

11 After the comment, add the following lines to get the map document's style gallery.

```
Dim pStyleGallery As IStyleGallery
Set pStyleGallery = pMxDoc.StyleGallery
```

12 Get an Enum of marker symbols from the Hazmat style's Hazmat category.

```
Dim pEnumMarkers As IEnumStyleGalleryItem
Set pEnumMarkers = pStyleGallery.Items _
        ("Marker Symbols", "Hazmat.style", "Hazmat")
```

13 Reset the Enum to be sure that the pointer points to the top of the list of marker symbols.

```
pEnumMarkers.Reset
```

14 Declare a variable to hold a style gallery item, and issue a Next to get the first marker symbol from the Enum.

```
Dim pStyleItem As IStyleGalleryItem
Set pStyleItem = pEnumMarkers.Next
```

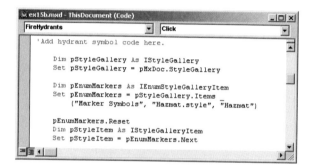

15 Declare a marker symbol variable.

```
Dim pMarker As IMarkerSymbol
```

You will set this variable when the loop you are about to write finds Fire Hydrant 4.

16 Start a loop to get each marker symbol in the Enum.

The loop will stop when there are no more items, because the last Next on an Enum returns Nothing.

```
Do Until pStyleItem Is Nothing

Loop
```

17 Inside the loop, add an If Then statement to check each marker symbol's name to see if it is Fire Hydrant 4.

```
If pStyleItem.Name = "Fire Hydrant 4" Then

End If
```

18 Inside the If Then statement, set the marker symbol variable and set the marker size to 14.

These two lines of code run only when the marker symbol's name is Fire Hydrant 4.

```
Set pMarker = pStyleItem.Item
pMarker.Size = 14
```

Setting layer symbols

Once the loop has found Fire Hydrant 4, you can stop it.

19 After setting the marker size, add Exit Do.

```
Exit Do
```

20 After the End If and before the Loop statement, add the following line to get the next marker symbol.

```
Set pStyleItem = pEnumMarkers.Next
```

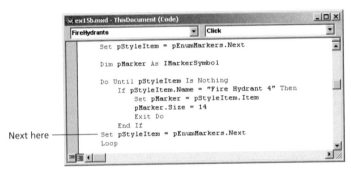

21 After the loop, create a simple renderer and set its Symbol property to the Fire Hydrant 4 marker symbol.

```
Dim pRenderer As ISimpleRenderer
Set pRenderer = New SimpleRenderer

Set pRenderer.Symbol = pMarker
```

As in the previous exercise, you will assign the renderer to the layer using the Renderer property on IGeoFeatureLayer. Again, the existing code in the click event has a pFlayer variable pointing to IFeatureLayer, so you can do QueryInterface.

22 Get the layer's IGeoFeatureLayer interface and set its Renderer property to the renderer you just created.

```
Dim pGFLayer As IGeoFeatureLayer
Set pGFLayer = pFLayer

Set pGFLayer.Renderer = pRenderer
```

When the map is zoomed out to the entire county, fire hydrants will draw in a messy blob. To see individual hydrants, you need to be zoomed in closer than 1:24,000. In the next step, you will set the layer's scale dependency.

23 Add one more line to set the layer's MinimumScale property.

```
pFLayer.MinimumScale = 24000
```

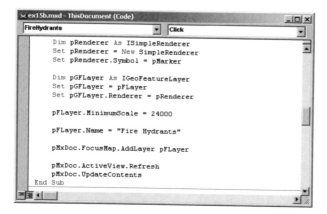

The code is ready to test.

24 Close Visual Basic Editor.

As in the previous exercise, you'll remove the layer that you added in step 2, then add it again.

25 In the ArcMap table of contents, right-click the Fire Hydrants layer and click Remove.

26 Click Add Layers and click Fire Hydrants.

The layer is added to the map, but it doesn't draw because the map's scale is zoomed out beyond 1:24,000.

Depending on your ArcMap window size, your scale may not match

27 In the Scale box, click the drop-down arrow and click 1:24,000.

The Fire Hydrants draw with the Fire Hydrant 4 symbol from the Hazmat style.

Scale is 1:24,000

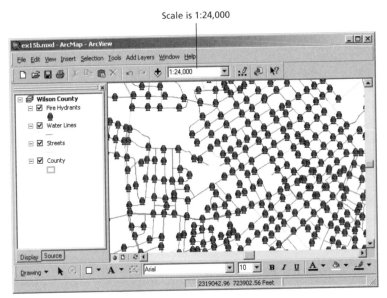

28 If you want to save your work, click the File menu in ArcMap and click Save As. Navigate to **C:\ArcObjects\Chapter15**. Rename the file **my_ex15b.mxd** and click Save. If you are continuing with the next exercise, leave ArcMap open. Otherwise close it.

Creating a class breaks renderer

When you map numeric attributes, such as population or income, you divide the range of values into classes and assign a different symbol to each class. (Here again, "class" just means a group of similar values, not an ArcObjects class.) From the user interface, you use the layer properties to set the number of classes, the value ranges, and the symbols for each class. As a programmer, you accomplish the same thing with a class breaks renderer.

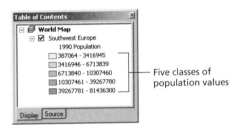 Five classes of population values

Each class in a legend has a range of beginning and ending values called break points. Records whose value falls within a class's break points belong to that class and are symbolized with that class's symbol.

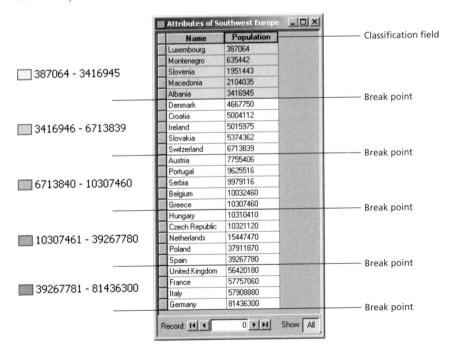

You can create a ClassBreaksRenderer with the code below.

```
Dim pCBR As IClassBreaksRenderer
Set pCBR = New ClassBreaksRenderer
```

IClassBreaksRenderer's Field property is used to specify an attribute to classify.

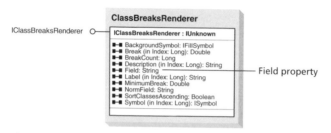

Field property

If you wanted to create classes of population values, for example, you would set the Field property equal to the Population field in the layer's attribute table.

```
pCBR.Field = "Population"
```

The BreakCount property sets the number of classes.

```
pCBR.BreakCount = 5
```

Each class is identified by an index position, starting with 0.

Index position   Break points are on the right

```
0  □ 387064 - 3416945
1  □ 3416946 - 6713839
2  □ 6713840 - 10307460
3  □ 10307461 - 39267780
4  ■ 39267781 - 81436300
```

You set class break points with the Break property. This property uses the class's index position as an argument. Any values that are smaller than the first break point (341645 in the code below) will be included in the first class. Any values larger than the last break point (81436300) will not be symbolized.

```
pCBR.Break(0)  = 341645
pCBR.Break(1)  = 6713839
pCBR.Break(2)  = 10307460
pCBR.Break(3)  = 39267780
pCBR.Break(4)  = 81436300
```

By default, the break point numbers are used to label each symbol. You can change the labels with IClassBreaksRenderer's Label property.

```
pCBR.Label(0) = "Very low"
pCBR.Label(1) = "Low"
pCBR.Label(2) = "Medium"
pCBR.Label(3) = "High"
pCBR.Label(4) = "Very high"
```

Like the Break property, Label uses index position numbers to set each class's label.

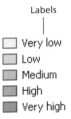

The last thing to do is set symbols for each class. In this chapter's first exercise, you made a blue line symbol. In the second exercise, you got a fire hydrant marker symbol from the style gallery. Now you will be symbolizing polygons, which use fill symbols. You can create a simple fill symbol with the code below.

```
Dim pFill As ISimpleFillSymbol
Set pFill = New SimpleFillSymbol
```

Say you wanted to make a blue color and assign it to the first class. You would make a new color object and set its blue property.

```
Dim pBlue As IRgbColor
Set pBlue = New RgbColor
pBlue.Blue = 255
```

Then you would assign the color to the fill symbol and assign the fill symbol to the renderer's first class at the 0 position.

```
pFill.Color = pBlue
pCBR.Symbol(0) = pFill
```

■ Very low
□ Low
▨ Medium
▨ High
▨ Very high

You would finish up by making four other colors and assigning them to the fill symbols of the other four classes.

13
14
15
16
17
18
19
20

Exercise 15c

During fires, floods, or similar emergencies, the county's emergency analysts make estimates of damage. In their work, they use a parcels layer in which parcels are classified as having low, medium, or high value.

In this exercise, you will modify the Parcels choice on the AddLayers menu. Currently, the layer draws in a single, random color. You will create a class breaks renderer to draw parcels in different colors depending on their value.

1 Start ArcMap and open **ex15c.mxd** in the **C:\ArcObjects\Chapter15** folder.

When ArcMap opens, you see the county boundary and a streams layer.

2 Click the Add Layers menu and click Parcels.

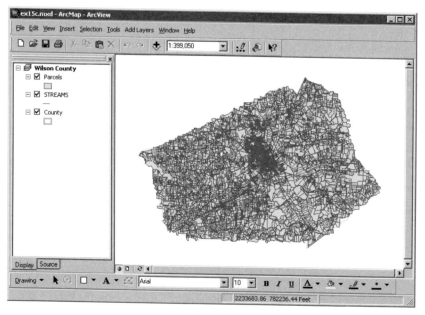

The parcels display in a random color.

3 In the table of contents, right-click the Parcels layer and click Open Attribute Table.

The third field in the table is called ParcelValue. Your code will symbolize the parcels according to the values in this field.

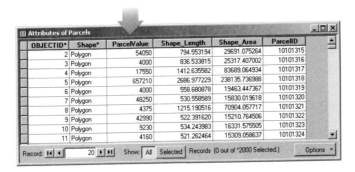

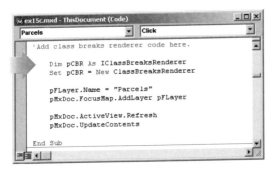

4 Close the attribute table.

5 Click the Tools menu and click Customize.

6 Click the Add Layers menu, right-click Parcels, and click View Source.

You see the Parcels click event. As in the previous exercises, the code to make the Parcels layer and associate it with a data source has already been written. The pFLayer variable, which is declared to IFeatureLayer, will again be useful for QueryInterface.

7 Scroll down to find the following commented line.

```
'Add class breaks renderer code here.
```

8 After the comment, add the following lines of code to create a class breaks renderer.

```
Dim pCBR As IClassBreaksRenderer
Set pCBR = New ClassBreaksRenderer
```

Creating a class breaks renderer

Next, you will specify the field you want to classify and how many class breaks you want.

9 Set the classification field to ParcelValue and the number of breaks to 3.

```
pCBR.Field = "ParcelValue"
pCBR.BreakCount = 3
```

You are creating three classes because you want to show high, medium, and low values.

10 Set the break point values.

```
pCBR.Break(0) = 50000
pCBR.Break(1) = 100000
pCBR.Break(2) = 9000000
```

The parcels range in value from $0 to over $8 million. The code here assigns parcel values between $0 and $50,000 to the first class, which has the index position of 0. Values between $50,001 and $100,000 are assigned to the second class, and values between $100,001 and $9 million to the third class.

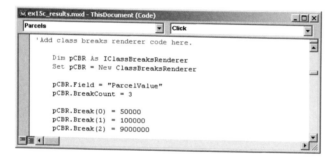

Next, you will set the renderer's Label property. If you don't set this property, the break values become the default labels. The Label property takes the same index position argument as the Break property.

11 Set the legend label values for the three classes.

```
pCBR.Label(0) = "Low"
pCBR.Label(1) = "Medium"
pCBR.Label(2) = "High"
```

In the next steps, you will create three shades of green for the renderer's three classes.

12 Make three RGB color objects.

```
Dim pGreenLight As IRgbColor
Dim pGreenMedium As IRgbColor
Dim pGreenDark As IRgbColor

Set pGreenLight = New RgbColor
Set pGreenMedium = New RgbColor
Set pGreenDark = New RgbColor
```

To assign color values to RgbColor objects, you usually set their Red, Green, and Blue color properties. That means writing nine more lines of code to make the three green colors. Once again, however, a shortcut is available. By using interface inheritance and a Visual Basic function, you will only need three lines of code.

As shown in the diagram below, the IColor Interface on the Color class has an RGB property. This property can be set with a VBA function also called RGB, which takes three color values as input and returns an OLE_COLOR object. (An OLE_COLOR is a number that represents a color's red, green, and blue values.) Specifying the three color values you want as arguments to a function allows you to put them all in the same line of code. For more information about the RGB function, see the Visual Basic help.

Since IRgbColor inherits from IColor (interface inheritance), you can use IColor's RGB property with the three IRgbColor variables you already have.

13 Set the colors' RGB property with the RGB function to make three shades of green.

```
pGreenLight.RGB = RGB(220,245,233)
pGreenMedium.RGB = RGB(118,168,130)
pGreenDark.RGB = RGB(34,102,51)
```

The class breaks renderer is almost complete. You have set its Break, BreakCount, Field, and Label properties. Next, you will set its Symbol property, which takes the same index position argument as the Break and Label properties.

14 Set the fill symbol for each class with a shade of green.

```
Dim pFill As ISimpleFillSymbol
Set pFill = New SimpleFillSymbol

pFill.Color = pGreenLight
pCBR.Symbol(0) = pFill

pFill.Color = pGreenMedium
pCBR.Symbol(1) = pFill

pFill.Color = pGreenDark
pCBR.Symbol(2) = pFill
```

15 Assign the renderer to the Parcels layer.

As in the previous exercises, you need to get the IGeoFeatureLayer interface to set its Renderer property. You will switch interfaces from IFeatureLayer.

```
Dim pGFLayer As IGeoFeatureLayer
Set pGFLayer = pFLayer

Set pGFLayer.Renderer = pCBR
```

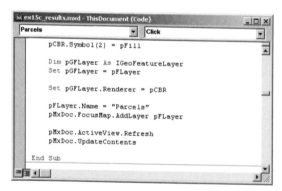

The code is ready to test.

16 Close Visual Basic Editor.

17 In the ArcMap table of contents, right-click the Parcels layer and click Remove.

18 Click Add Layers and click Parcels.

The layer is added to the map, and the parcels draw in three shades of green. In the table of contents, you see their legend and labels.

19 Click the View menu, point to Bookmarks, and click Downtown Wilson City.

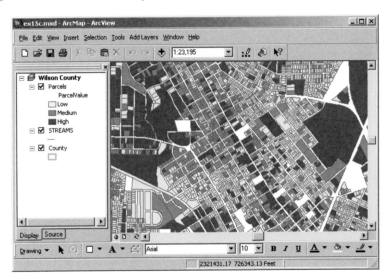

By overlaying layers of fire or flood data on the parcels, analysts can get a quick sense of how severe the property damage is likely to be.

20 If you want to save your work, click the File menu in ArcMap and click Save As. Navigate to **C:\ArcObjects\Chapter15**. Rename the file **my_ex15c.mxd** and click Save. If you are continuing with the next chapter, leave ArcMap open. Otherwise close it.

Using ArcCatalog objects in ArcMap

Adding layer files to ArcMap
Making your own Add Data dialog box

Up to this point, you haven't worked with any ArcCatalog objects. Although you won't customize ArcCatalog in this book, you will learn how to program some of its objects in ArcMap.

No matter which ArcGIS application you are working in (ArcMap, ArcCatalog, or ArcScene™, if you have the ArcView 3D Analyst™ extension), you have access to all the ArcObjects classes. Just because classes appear on a certain diagram doesn't mean they are limited to use in a particular application. Diagrams are just convenient, more or less logical, groupings of the thousands of different ArcObjects classes and interfaces. Theoretically, there could be one huge, wall-sized diagram called ArcObjects.

ArcCatalog and ArcMap have similar starting points. In the diagram below, you see that the ArcCatalog application is composed of a document called GxDocument. Just like ArcMap, ArcCatalog has two predefined variables, Application and ThisDocument, that refer to the application and the document.

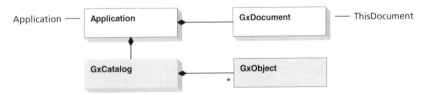

Despite this similarity, there are some differences. For one thing, customizations in ArcMap can be made to a map document, to a base template (a class of map documents), or to the normal template (the class of all map documents). That means that the ThisDocument variable in ArcMap may refer to someMap.mxd, someTemplate.mxt, or normal.mxt, depending on where you are saving your customizations.

ArcCatalog, on the other hand, doesn't have documents or base templates. All it has is a normal template. Any customizations you make are applied whenever the application is opened. That means that the ThisDocument variable in ArcCatalog always refers to its normal template, normal.gxt.

If you look at the tree view (the left-hand pane) in ArcCatalog, you'll see that it consists of files and folders organized below an object called Catalog. This structure is reflected in the other part of the diagram above. It shows that the ArcCatalog application is also composed of a GxCatalog object, which in turn is composed of many GxObjects. A GxObject is any file, folder, disk connection, or other object you can click in the tree view.

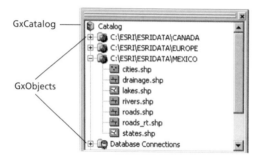

All ArcCatalog objects that can be displayed in the tree view share the abstract GxObject class. The IGxObject interface on this class has properties and methods for getting basic information about objects, like their names.

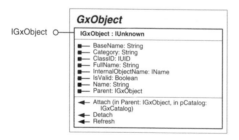

By now, you have probably noticed that many ArcCatalog objects have a Gx prefix, just as many ArcMap objects have an Mx prefix. These prefixes don't mean anything special—they just save you some typing. It's easier to use names like MxDocument, GxDocument, and GxObject than it would be to use names like ArcMapDocument, ArcCatalogDocument, and ArcCatalogObject.

The simplified diagram below shows that one of the types of GxObject is a GxFile. GxFile is a coclass and so are its subclasses. GxFiles represent the various types of files that ArcMap or ArcCatalog can create.

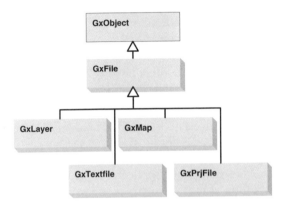

In the first exercise of this chapter, you will work with one of these subclasses—the GxLayer class, which represents layer files (files with the .lyr extension). You will write code to make a GxLayer object and add it to ArcMap.

In addition to various GxObject classes, the ArcCatalog diagram also contains five coclasses that represent dialog boxes. These coclasses, shown below, are not connected to other classes and are not hierarchically related. They stand by themselves.

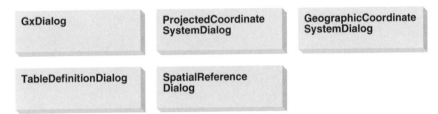

The dialog boxes that you make from these coclasses are forms. They are like the VBA forms you worked with early in this book, but they have been given special properties and methods by ESRI. TableDefinitionDialog, for example, is a predefined form with input boxes that a user fills in to make a table. GxDialog is the Add Data dialog box that you use to add layers to ArcMap. You will use GxDialog in the second exercise to make a customized Add Data dialog box.

In chapter 10, you learned that all the ArcObjects classes are COM classes, and you learned about some of the advantages of COM. An advantage that was not mentioned is that COM classes are not tied into any one application—they can be used in any COM application.

That means that you can use ArcObjects outside of ESRI applications. Say you want to add mapping functionality to a word-processing document. If you have Microsoft Word—also a COM-based application with built-in VBA—you can do it. Open a new Word document, use its Customize dialog box to make a button, open Visual Basic Editor to the button's click event procedure, make a reference to ESRI ArcObjects, and write VBA code using ArcObjects. (When you are programming in Word, the predefined Application and ThisDocument variables refer to the Word application and its document, not to the ArcMap application and a map document.) You can also use ArcObjects in other Microsoft applications, like Microsoft Excel, Microsoft PowerPoint®, and Microsoft Access.

The reverse is also true. Those applications all have object model diagrams of their classes, and you can use their classes in ESRI applications. To see how it's done, go to arcobjectsonline.esri.com. Type **spell check** into the search box. There you'll see sample VBA code that gets the Microsoft Word application and uses its spell checker inside ArcMap to check the spelling of text elements on a layout page.

The VBA coding techniques you have learned in this book are the same in any COM application that uses VBA. To program in Microsoft Word, you don't have to relearn any syntax or coding techniques. All you have to do is look at Word's object model diagram to see what classes are available and what their properties and methods are.

Adding layer files to ArcMap

As you know from chapter 15, when a layer is added to ArcMap, it is assigned a default symbol and a random color. A user might go on to symbolize the layer in some way and then save the map. Saving the map saves both the layer's symbology and the path to its data set into the current .mxd file. When the user closes and reopens the map document, the layer is there, displaying its data with the symbol and color previously assigned.

Suppose a coworker sees that layer and wants to use it in their own map document. How can you move a layer to another map document and preserve its symbology? You do it by making a layer file from the ArcMap user interface. A layer file, which has the extension .lyr, is like a miniature map document. It stores information specific to a single layer—its symbology, the path to its data set, and any other information that can be set in the Layer Properties dialog box. Layer files are portable and can be added to other map documents, where they look just the same as they did in the original.

In chapter 14, you went through the process of creating a new feature layer, getting a data set for it, associating the layer with the data, and adding the layer to ArcMap. When you add a layer file to ArcMap, you get to skip the first three steps. A layer file already *is* a layer, it already knows the path to its data set, and it's already associated with the data. As a programmer, all you have to do is get the file from its location on disk and add it to ArcMap.

The following diagram shows that a GxLayer is a type of GxFile and that both are GxObjects:

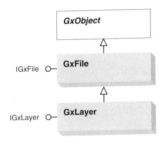

Since both GxFile and GxLayer are coclasses, you can create either one directly.

```
Dim pGxLayer As IGxLayer
Set pGxLayer = New GxLayer
```

To get a layer file from disk, you need the IGxFile interface's Path property, which is a string consisting of a path.

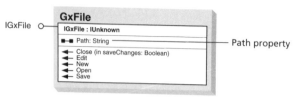

IGxFile

Path property

Since GxLayer is a type of GxFile, it has all the interfaces of GxFile. Therefore, you can switch interfaces from IGxLayer to IGxFile.

```
Dim pGxFile As IGxFile
Set pGxFile = pGxLayer
```

You then set the Path property to the location you need.

```
pGxFile.Path = "C:\arcobjects\MyRiversLayer.lyr"
```

You add the layer to ArcMap with the AddLayer method on IMxDocument or IMap. AddLayer takes an ILayer object as its argument. To get the layer file's ILayer interface, you could declare an ILayer variable and set it with the GxLayer's Layer property.

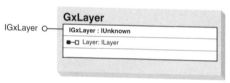

IGxLayer

Even more simply, you could use chaining and write the following single line of code:

```
pMxDoc.AddLayer pGxLayer.Layer
```

You can do it that way because the Layer property returns ILayer, which is exactly what the AddLayer method needs.

As you know, after adding a layer you have to refresh the table of contents and the active view.

Exercise 16a

In this exercise, you will write code that allows the emergency response analysts to add a parcels layer file (*.lyr) to ArcMap. You will begin by making the layer file yourself from the ArcMap user interface.

You may wonder why you, as a programmer, should work from the user interface. It's because some of your tasks can be accomplished without any programming. When you can make something you need from the user interface, it saves you time because you don't have to write, update, or fix any code.

You will symbolize the parcels layer with the Sahara Sand color, turn its map tips on, and label it with parcel identification numbers. Since there are thousands of parcels, you would never be able to see all the IDs at once. You'll set them to display at scales larger than 1:4,000. If a user is zoomed out beyond that scale, they can still use map tips to see the IDs of individual parcels.

1 Start ArcMap and open **ex16a.mxd** in the **C:\ArcObjects\Chapter16** folder.

You see the Parcels layer for Wilson County.

2 In the table of contents, right-click the Parcels symbol. In the color palette, click Sahara Sand.

In the following steps, you will set the other properties for this layer: map tips, labels, and the label display scale.

3 In the table of contents, right-click on the Parcels layer and click Properties to open the Layer Properties dialog box.

4 Click the Display tab. Check the Show MapTips box.

The attribute you want to show up as a map tip has to be defined as the primary display field. You make this setting on the Fields tab.

5 Click the Fields tab. In the drop-down list, click ParcelID.

The map tips are now set to show Parcel ID numbers.

6 Click the Labels tab. In the Label Field drop-down list, click ParcelID.

The parcels will also be labeled with their ID numbers.

7 At the bottom of the tab, click the Scale Range button to open the Scale Range dialog box.

8 Click the "Don't show labels when zoomed" option. In the "Out beyond:" box, type **4000**. Click OK.

9 At the top of the Labels tab, check the box to label features in this layer. Click OK.

The parcels display in the Sahara Sand color, but you don't see any labels because you are zoomed too far out.

10 Hover your mouse pointer over any parcel to see its parcel ID in a map tip.

Now you will zoom in to display the labels.

11 Click the View menu, point to Bookmarks, and click Downtown Wilson City. (If your map is still zoomed out too far, set the map scale to **1:4000** on the Standard toolbar and press Enter.)

You see the parcels labeled with their ID numbers. This is how you want the layer to look, so you will now make a layer file from it.

12 In the table of contents, right-click the Parcels layer and click Save As Layer File. Navigate to **C:\ArcObjects\Chapter16** (or the location where you installed the data for this book). On the Save Layer dialog box, click the New Folder button.

New Folder

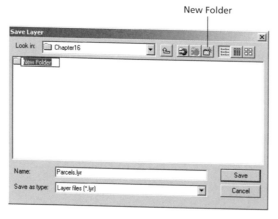

You are going to save the layer file into its own folder.

13 Rename the new folder **Layers** and press Enter. Double-click the Layers folder to open it.

14 At the bottom of the dialog, in the Name box, rename the layer file **ParcelIDs.lyr**. Click Save.

Now you can write the code to add this layer file to ArcMap.

15 Open the Customize dialog box.

16 Click the Add Layers menu, right-click Parcel IDs, and click View Source.

You will write your code in the ParcelIDs click event procedure. The first thing you need to do is make a GxLayer object.

17 In the click event procedure, create a new GxLayer object that points to the IGxLayer interface.

```
Dim pGxLayer As IGxLayer
Set pGxLayer = New GxLayer
```

You will use this variable later to add the layer to ArcMap. Right now, you will switch interfaces to IGxFile, which has the Path property you need to get the layer.

18 Declare an IGxFile variable and set it equal to pGxLayer.

```
Dim pGxFile As IGxFile
Set pGxFile = pGxLayer
```

19 Set the layer file's Path property. (If you saved the layer file to a different location, use your path here.)

```
pGxFile.Path = _
    "C:\arcobjects\chapter16\layers\parcelids.lyr"
```

The layer can now be added to the map.

20 Declare and set a variable to the IMxDocument interface.

```
Dim pMxDoc As IMxDocument
Set pMxDoc = ThisDocument
```

IMxDocument has the AddLayer method.

21 Write a line of code to add the layer to the map.

```
pMxDoc.AddLayer pGxLayer.Layer
```

22 Add two lines of code to redraw the table of contents and the map display area.

```
pMxDoc.ActiveView.Refresh
pMxDoc.UpdateContents
```

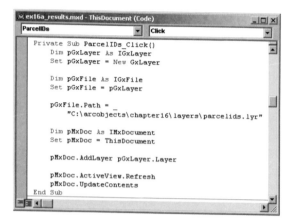

```
ex16a_results.mxd - ThisDocument (Code)
ParcelIDs                          Click

   Private Sub ParcelIDs_Click()
      Dim pGxLayer As IGxLayer
      Set pGxLayer = New GxLayer

      Dim pGxFile As IGxFile
      Set pGxFile = pGxLayer

      pGxFile.Path = _
         "C:\arcobjects\chapter16\layers\parcelids.lyr"

      Dim pMxDoc As IMxDocument
      Set pMxDoc = ThisDocument

      pMxDoc.AddLayer pGxLayer.Layer

      pMxDoc.ActiveView.Refresh
      pMxDoc.UpdateContents
   End Sub
```

Your code is finished.

23 Close Visual Basic Editor.

Before testing the menu choice, you will zoom out to the full extent of the parcel layer and then delete the original Parcels layer.

24 On the Tools toolbar, click the Full Extent button. In the table of contents, right-click Parcels and click Remove.

25 In ArcMap, click the Add Layers menu and click Parcel IDs.

The parcel layer draws but you don't see any ID numbers, because you are zoomed too far out. However, the layer's map tips are turned on.

26 Hover your mouse over any parcel to see its ID in a map tip.

27 Click the View menu, point to Bookmarks, and click Downtown Wilson City.

The parcel ID labels display (provided your scale is greater than 1:4,000).

28 If you want to save your work, click the File menu in ArcMap and click Save As. Navigate to **C:\ArcObjects\Chapter16**. Rename the file **my_ex16a.mxd** and click Save. If you are continuing with the next exercise, leave ArcMap open. Otherwise close it.

Making your own Add Data dialog box

Since chapter 14, you have written code to add different types of layers to ArcMap. In each case, your code has contained a path to data. In some emergencies, however, the data is updated so often that you can't code the paths in advance. A toxic plume, for instance, may be represented by dozens of layers as it changes its direction, height, size, speed, and concentration.

In situations like these, users will have to rely on the Add Data dialog box to navigate to the latest data. You can make their job easier, however, with a little customization. For example, you can make the Add Data dialog box open to a specific folder or set it to display only a certain file type, such as .lyr files.

The Add Data dialog box is an ArcCatalog object called a GxDialog. Basically, it is a form that displays ArcCatalog's tree view inside ArcMap. It has also been coded by ESRI to take user-selected data sets and add them as layers to ArcMap.

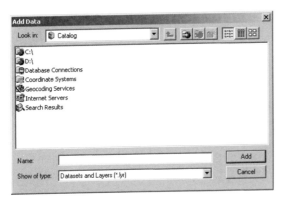

As shown below, GxDialog is a coclass with properties that you can set including AllowMultiSelect, ButtonCaption, Name, ObjectFile, RememberLocation, StartingLocation, and Title. You can customize the dialog box within the limits of these properties. You create one with the following code:

```
Dim pGxDialog As IGxDialog
Set pGxDialog = New GxDialog
```

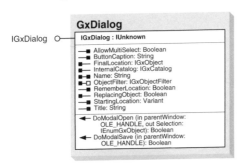

Setting the ButtonCaption, StartingLocation, and Title properties with the next three lines of code produces the dialog box that follows. The dialog box opens at Catalog (the top level).

```
pGxDialog.ButtonCaption = "Add"
pGxDialog.StartingLocation = "Catalog"
pGxDialog.Title = "Add Data"
```

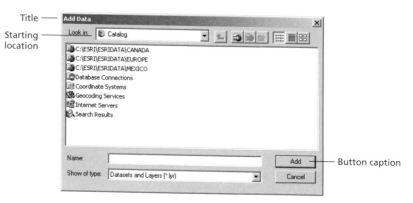

Normally, the user can select multiple objects in the dialog box by holding down the Shift or Control keys. This is because the AllowMultiSelect property is set to True.

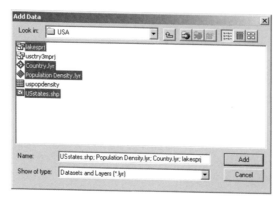

You can restrict the user to a single selection by setting this property to False.

```
pGxDialog.AllowMultiSelect = False
```

The dialog box above displays coverages, layer files, a raster data set, and a shapefile. What if you want the user to see only raster data sets? The GxObjectFilter class, associated with GxDialog, allows you to control the types of data that are displayed.

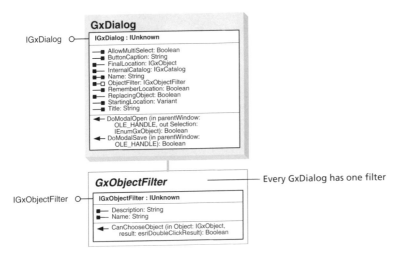

GxObjectFilter is an abstract class with more than thirty different filters, each of which is a coclass.

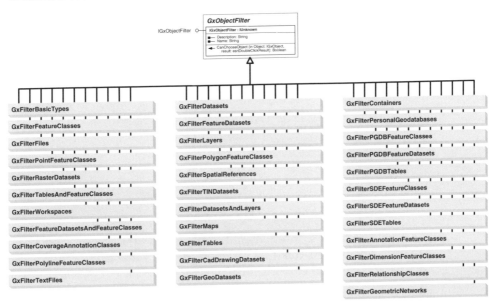

To make the Add Data dialog box display only raster data sets, you would create a raster data set filter (GxFilterRasterDatasets). Similarly, to display only layer files, you would create a layer file filter with the following code:

```
Dim pLFilter As IGxFilterLayers
Set pLFilter = New GxFilterLayers
```

You associate a filter with a GxDialog using IGxDialog's ObjectFilter property. This is a byRef property and requires the Set keyword.

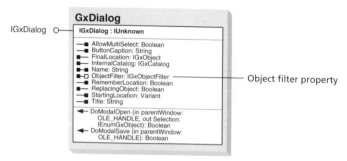

The following line of code sets the ObjectFilter property:

```
Set pGxDialog.ObjectFilter = pLFilter
```

When the dialog box opens, the user selects objects from the tree view. When the user clicks OK, the dialog box closes. Any selected objects are stored in an Enum, like the one you used in the last chapter to hold style gallery items.

In this case, the Enum is called an EnumGxObject. Its IEnumGxObject interface has the familiar Next and Reset methods, which you might use in a loop to get individual objects. Since the dialog box displays the ArcCatalog tree view, the objects that can be selected to fill the Enum are all GxObjects and have the IGxObject interface.

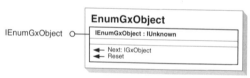

Before you code the GxDialog to open, you declare a variable for the Enum.

```
Dim pLayerFiles As IEnumGxObject
```

You would normally write another line of code to set this variable, but EnumGxObject is a little different. As you'll see in a moment, the variable is set for you when you (or your users) open and close the dialog box.

To open a GxDialog, you run the DoModalOpen method on the IGxDialog interface. (This is a little different from the Show method you used to open a simple VBA form in chapter 4.) DoModalOpen has two arguments. The first is a window handle, which you can set to 0, as you did in chapter 14. The second is the Enum variable that will hold the selected layer files.

```
pGxDialog.DoModalOpen 0, pLayerFiles
```

The DoModalOpen method has been coded by ESRI to set the Enum variable for you. When the method runs, it will not only open the dialog box, but also set pLayerFiles equal to the collection of GxObjects that the user selects.

To get individual layer files out of the Enum, you use the Next method. Since the Next method returns IGxObject, you declare an IGxObject variable and set it with the Next method.

```
Dim pLayerFile As IGxObject
Set pLayerFile = pLayerFiles.Next
```

In the previous chapter, you wrote a looping statement to process each item in the Enum. In this exercise, you are going to restrict the user to selecting a single object, so you won't need a loop.

Exercise 16b

In exercise 14b, you wrote code to add an air photo to a map. At that time, a train had crashed and a broken car was emitting a plume of toxic smoke. Now the plume's movement is being monitored with new air photos taken every hour. The photos are used to digitize the plume's outline into polygon feature classes. These feature classes are then symbolized and saved as layer (.lyr) files.

In this exercise, you will create a GxDialog that is customized to save time for the emergency analysts who work with the plume layer files. The dialog will open at the location where the layer files are stored, and will display only .lyr files. It will also restrict the user to making a single selection from the dialog box.

1 Start ArcMap and open **ex16b.mxd** in the **C:\ArcObjects\Chapter16** folder.

The map shows layers of schools, fire stations, and railroads, as well as a graphic marking the train crash site.

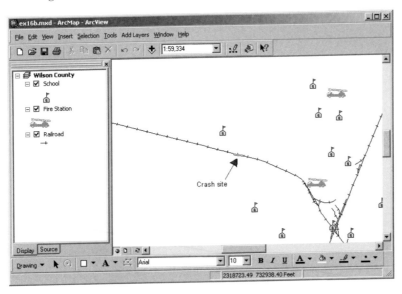

2 Open the Customize dialog box.

3 Click the Add Layers menu, right-click Toxic Layers, and click View Source.

4 Inside the Toxic Layers click event procedure, create a new GxDialog object.

```
Dim pGxDialog As IGxDialog
Set pGxDialog = New GxDialog
```

In the following steps, you will customize the dialog box by setting its properties.

5 Set the dialog box's title to **Add Toxic Layer**.

```
pGxDialog.Title = "Add Toxic Layer"
```

6 Set the button's caption to **Add Layer**.

```
pGxDialog.ButtonCaption = "Add Layer"
```

7 Set the AllowMultiSelect property to False.

```
pGxDialog.AllowMultiSelect = False
```

8 Set the dialog box's StartingLocation property. (If you installed the data for this book in a different location, type the correct path.)

```
pGxDialog.StartingLocation = _
    "C:\arcobjects\data\wilson_nc"
```

Since every toxic plume is represented by a layer file, you don't need to show any other types of data in the dialog box. You will create a new object filter to filter out everything except .lyr files.

9 Declare and set a variable to create a new object filter.

```
Dim pGxFilter As IGxObjectFilter
Set pGxFilter = New GxFilterLayers
```

10 Set the dialog box's ObjectFilter property equal to the filter.

```
Set pGxDialog.ObjectFilter = pGxFilter
```

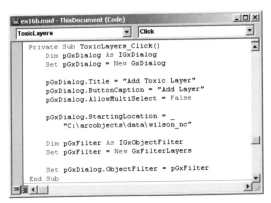

Before running the method to open the dialog box, you need to declare an Enum variable. This variable will hold the layer files that the user selects. (In this case, because AllowMultiSelect is False, the Enum can only contain one object.)

11 Declare an Enum variable to hold the user's selected layer file.

```
Dim pLayerFiles As IEnumGxObject
```

12 Add a line of code to open the dialog box.

```
pGxDialog.DoModalOpen 0, pLayerFiles
```

When this line runs, the dialog box opens and the code pauses until the user makes a selection. When the user clicks the Add Layer button, the pLayerFiles variable is set automatically, and the procedure resumes.

What happens if the user clicks the Cancel button instead of selecting a layer file? In that case, the Enum will be empty. To deal with this possibility, you'll write an If Then statement. You'll declare and set a variable to hold the layer file that you expect the Enum to contain. If the Enum is empty, however, you'll exit the procedure. (Enums are usually used with looping statements, but in this case, all you need to know is whether there is an object in the Enum or not.)

13 Declare and set a variable to hold a GxObject. Use the Next method on the Enum to get its first object.

```
Dim pLayerFile As IGxObject
Set pLayerFile = pLayerFiles.Next
```

14 Use an If Then statement to test whether pLayerFile is Nothing. If it is Nothing (the Enum is empty), exit the procedure.

```
If pLayerFile Is Nothing Then
    Exit Sub
End If
```

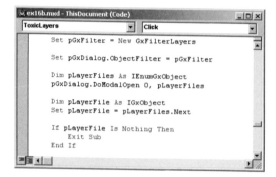

As long as the Enum is not empty, you can safely assume that the user has selected a layer file, and that the pLayerFile variable is now holding it.

Now you want to add the layer file to ArcMap. The process is the same one that you used in the last exercise. You set a variable to the IGxLayer interface so you can then get its Layer property, which returns ILayer. You can then run the AddLayer method on IMxDocument, using ILayer as its argument.

Since the pLayerFile variable is pointing to IGxObject, and you need IGxLayer, you have to switch interfaces.

15 Declare a variable to IGxLayer and set it equal to the layer file variable you already have.

```
Dim pGxLayer As IGxLayer
Set pGxLayer = pLayerFile
```

16 Create an IMxDocument variable.

```
Dim pMxDoc As IMxDocument
Set pMxDoc = ThisDocument
```

17 Add the layer to the active map.

```
pMxDoc.AddLayer pGxLayer.Layer
```

18 Refresh the table of contents and map display.

```
pMxDoc.ActiveView.Refresh
pMxDoc.UpdateContents
```

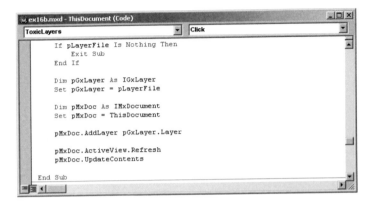

The code is ready to test.

19 Close Visual Basic Editor.

20 Click the Add Layers menu and click Toxic Layers.

The Add Toxic Layer dialog box opens to the Wilson_NC data folder, and only plume layer files are available. There are four layer files that were created at one-hour intervals beginning at 1 P.M.

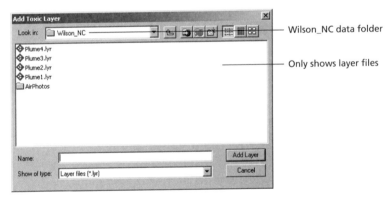

21 In the Add Toxic Layer dialog box, click Plume4.lyr.

If you like, try holding down the Shift or Control keys and clicking another layer. You can select only one at a time.

22 With Plume4.lyr selected, click Add Layer.

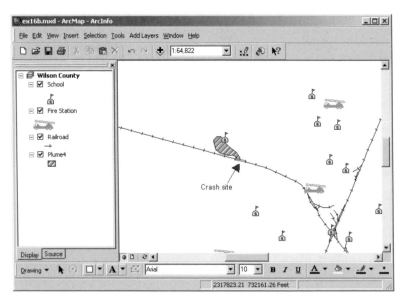

The layer for Plume4 is added to the map. You see that it almost covers the school. Next, you will test the Cancel button.

23 Click the Add Layers menu and click Toxic Layers.

24 Click Cancel.

The dialog box disappears and no layers are added to the map.

25 If you want to save your work, click the File menu in ArcMap and click Save As. Navigate to **C:\ArcObjects\Chapter16**. Rename the file **my_ex16b.mxd** and click Save. If you are continuing with the next chapter, leave ArcMap open. Otherwise close it.

Controlling feature display

Making definition queries
Selecting features and setting the selection color

In this and the next two chapters, you will cover some old ground from early in the book and also learn a lot of new things as you code an application that displays toxic waste sites for selected U.S. states. The application will consist of two maps, or data frames: one will be an overview map of the United States and the other will be a detail map of a selected state. When the user picks a state from a drop-down list, the selected state will highlight on the overview map. On the detail map, only counties and EPA-designated toxic waste sites for the selected state will draw—all other features (except for some background layers) will be filtered out. The detail map will zoom to the selected state and the map's title will change to reflect the selection.

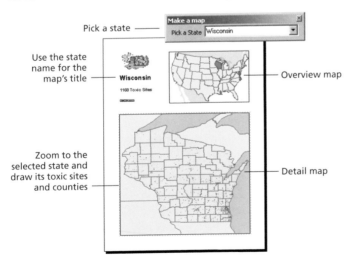

Pick a state

Use the state name for the map's title

Zoom to the selected state and draw its toxic sites and counties

Overview map

Detail map

Some of the work has already been done for you. A toolbar for the application has been created and a combo box has been populated with U.S. state names. (You wrote similar code to populate a combo box in chapter 7.) In this chapter, you'll concentrate on code to do two related tasks: create a definition query for the detail map, and select a feature for the overview map.

The code for these two tasks uses different interfaces and properties and methods, but conceptually they have one important thing in common: they are both based on queries.

A common type of GIS question is "Which features meet such-and-such criteria?" To get an answer, you have to phrase your question using the special syntax of a query. For example, if the question is "Which states in the United States have a population over twelve million?", the query statement would be:

```
"State_Population > 12000000"
```

A query statement is a text string made up of a field name, an operator, and a value. In the example above, State_Population is a field in a layer attribute table, the greater than sign (>) is the operator, and 12000000 is the value. ArcGIS users and programmers alike use query statements to make definition queries and to select features. In a definition query, only those features that meet the query statement's criteria are displayed. In a feature selection, features that meet the criteria are high-lighted in a special color (a red outline in the graphic below).

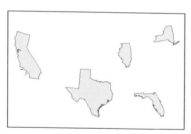

Definition query

Selected features

In this chapter, you will learn how to make these two types of selections. In chapter 18, you will perform operations on the selected features, such as counting them and zooming to them. In chapter 19, you will get attribute information from selected features and use that information to title the map.

A major challenge in coding this application will be working with two maps that have to be kept in synch: whenever the user selects a state, the overview map will redraw to highlight that state and the detail map will redraw so that only that state's features display. When the user clears the selection, it will have to be cleared in both maps. This means that you'll have to juggle pairs of variables and pay close attention to the order in which your lines of code execute.

You will write your code in an event procedure that you haven't used before: a SelectionChange event. Just as buttons have a click event, and tools have MouseUp, MouseDown, and MouseMove events, so combo boxes have a SelectionChange event. Code in a combo box's SelectionChange event procedure runs whenever a user makes a new selection in the combo box.

Pick a State combo box ——

Making definition queries

In this exercise, you will write code for the detail map that shows counties and toxic waste sites. You'll code the overview map in the next exercise. Since the code for both maps goes in the same SelectionChange event procedure, in this exercise you will set up variables to work with both maps.

In previous chapters, you've worked with map documents that contain only a single data frame, which you could get with the FocusMap property on IMxDocument. In this application, you need to get two maps. IMxDocument has another property, the Maps property, that lets you do this. The Maps property returns the IMaps interface of a map collection object. A map collection is like the Enums you've worked with in the last two chapters—it's a list of objects. The IMaps interface has properties and methods to work with each map in the collection.

Assuming that you already have a pMxDoc variable referencing ThisDocument, the following code gets the IMaps interface of the map collection:

```
Dim pMaps As IMaps
Set pMaps = pMxDoc.Maps
```

You can learn more about the IMaps interface and its properties and methods by reading about IMap in the ArcObjects developer help (shown below).

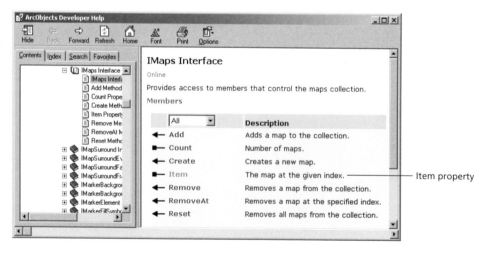

IMaps doesn't have a Next method to move through the map collection, but each map in the collection does have an index position number. If you know a map's index number, you can get the map with the Item property on IMaps. This property returns a map's IMap interface given its index position number. (IMaps and IMap are two different interfaces.)

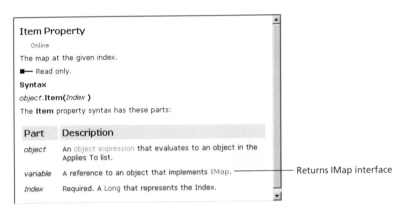

Item Property

Online

The map at the given index.

■— Read only.

Syntax

object.**Item**(*Index*)

The **Item** property syntax has these parts:

Part	Description
object	An object expression that evaluates to an object in the Applies To list.
variable	A reference to an object that implements IMap. ——— Returns IMap interface
Index	Required. A Long that represents the Index.

As shown below, the maps are numbered starting with 0. The EPA map at the top of the table of contents is at position 0, and the USA map is at position 1.

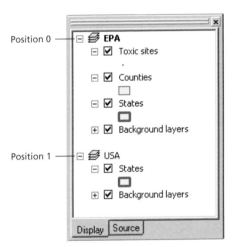

Position 0

Position 1

To get the EPA map's IMap interface, you would write the following code using the Item property:

```
Dim pEPAMap as IMap
Set pEPAMap = pMaps.Item(0)
```

Once you have a variable referring to a map, you can get specific layers from it with the Layer property on IMap. In fact, you have already done this (in chapter 11), so you may recall that layers have index position numbers just as maps do. In the graphic above, the States layer is at position 2 in the EPA map. To get this layer, you would write the following code:

```
Dim pStatesLayer As ILayer
Set pStatesLayer = pEPAMap.Layer(2)
```

Once you have a layer, you can make a definition query. You make a definition query by setting the DefinitionExpression property on the IFeatureLayerDefinition interface. The property is set to a string that represents a query statement.

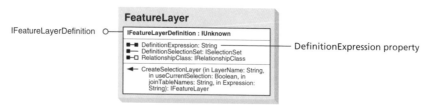

IFeatureLayerDefinition is on the FeatureLayer class, which inherits from the abstract Layer class. Since you already have a variable pointing to ILayer, you can switch interfaces.

```
Dim pStatesLayerDef As IFeatureLayerDefinition
Set pStatesLayerDef = pStatesLayer
```

Say that you wanted to display only those states with a population greater than twelve million. You would set the DefinitionExpression property equal to the query statement string that you saw in this chapter's introduction.

```
pStateLayerDef.DefinitionExpression _
    "State_Population > 12000000"
```

Or say that you wanted to display just the state of Arizona. Then you would set the DefinitionExpression property as follows:

```
pStateLayerDef.DefinitionExpression _
    "State_Name = 'Arizona'"
```

The entire query statement is double quoted because it's a string. Arizona is single quoted because it's a string within a string.

In the application you are developing, users pick a state from a combo box. You don't know in advance which state will be picked, so the query statement string will use a variable for the state name.

Combo boxes have an EditText property that returns whatever value the user selects as a string. In the graphic above, Hawaii is selected. The second line of code below returns Hawaii as a string and assigns it to the strState variable.

```
Dim strState As String
strState = cboStateNames.EditText
```

(You may remember from the early chapters of this book that when you declare an intrinsic variable, like a string, you don't declare it to an interface or use the Set keyword to set it.)

Your query statement's search value, then, is stored in the strState variable. Since this variable represents a state name, you might expect to put single quotes around it, as in the Arizona example. In fact, however, this doesn't work because VBA interprets anything inside quotation marks as a literal string.

Say you set a variable X equal to the string "Hello". If you then attempt to display "Hello" in a MsgBox with the line of code

```
MsgBox "X"
```

you will get the result shown below.

The solution is to break the query statement into three separate strings and concatenate them. That way, VBA can evaluate the variable and the result can end up being single quoted (since it's a string inside another string).

The first part of the query consists of the field name, the operator, and the first single quote:

```
"State_Name = '"
```

Then comes the variable containing the state name string:

```
strState
```

The last part of the query is a single quote surrounded by double quotes:

```
"'"
```

You will form the query statement by concatenating (&) these three strings.

```
"State_Name = '" & strState & "'"
```

When VBA reads the line of code above, the strState variable is evaluated first to get the state name string, which is then concatenated with the first and last parts of the query to make the full query statement.

Since this query statement is pretty long and your code is going to use it in a couple of different places, it will be convenient to set the entire string equal to a variable and use the variable instead.

```
Dim strQuery As String
strQuery = "State_Name = '" & strState & "'"
```

When you set the DefinitionExpression properties for the Toxic sites and Counties layers equal to this variable, only the features for the state picked from the combo box will draw; the others will be filtered out.

```
pStateLayerDef.DefinitionExpression = strQuery
```

Every time the user picks a new state from the combo box, the SelectionChange event procedure runs again, and the layers' DefinitionExpression properties are reset. Your code also has to be able to remove a layer's DefinitionExpression property, so that all features from the layer can be displayed again. When the user wants the toxic sites for all states to appear in the detail map, they pick <Show All> in the combo box. Your code handles this choice by setting the DefinitionExpression property to a blank string.

```
pStateLayerDef.DefinitionExpression = " "
```

You can write an If Then statement to test whether the user has picked a state or <Show All>. If they pick a state, you set the definition expression to strQuery. If they pick <Show All>, you set the expression equal to blank quotes.

There is one more thing to do. After the If Then statement runs, you need to refresh the EPA map. You could do this by running the Refresh method on IActiveView (which you get with the ActiveView property on IMxDocument). However, when the user is working in layout view, as in this exercise, the Refresh method refreshes the entire layout page—both maps (data frames) as well as other elements on the layout.

In this case, you have made a change to just one map and that's all you want to refresh. You can do this because maps also have the IActiveView interface.

With a variable pointing to IMap, you can do QueryInterface to IActiveView and then run the Refresh method on a particular map.

```
Dim pEPAActiveView As IActiveView
Set pEPAActiveView = pEPAMap

pEPAActiveView.Refresh
```

Exercise 17a

Your company cleans up toxic sites in the contiguous United States. You have a database of more than fourteen thousand such sites as defined by the U.S. Environmental Protection Agency (EPA). Project managers at your company want to be able to pick a U.S. state and immediately see a map of it and its toxic sites.

Over the next three chapters, you will write code for the Pick a State combo box shown below on the Make a map toolbar. With this toolbar, a manager will be able to select a state and get a detailed map of its toxic sites. An overview map will highlight the state, and the map's title information will reflect the state's name, the number of toxic sites it has, and the current date.

In this exercise, you will write code to determine which U.S. state is selected in the Pick a State combo box. Using the state's name, you'll create a query statement to set DefinitionExpression properties for the toxic sites and counties within the state. When finished, you will be able to pick a state and see its toxic sites and county outlines in the detail map.

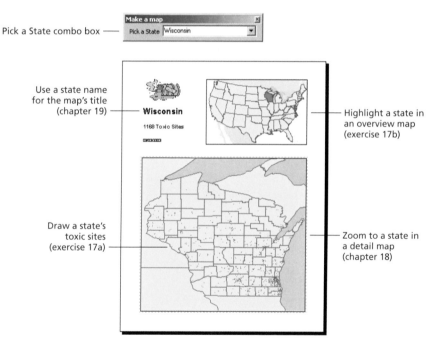

Pick a State combo box

Use a state name for the map's title (chapter 19)

Highlight a state in an overview map (exercise 17b)

Draw a state's toxic sites (exercise 17a)

Zoom to a state in a detail map (chapter 18)

1 Start ArcMap and open **ex17a.mxd** in the **C:\ArcObjects\Chapter17** folder.

The map opens in layout view. The USA map displays in the top data frame while the EPA map displays in the bottom frame.

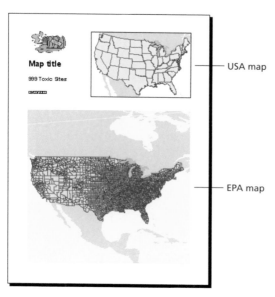

— USA map

— EPA map

2 On the Make a map toolbar, click the Pick a State drop-down arrow to see state names.

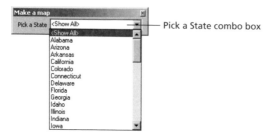

— Pick a State combo box

The code that adds the <Show All> choice and the state names to the combo box list has been provided for you. Next, you will write code to make it work.

3 Open the Customize dialog box.

4 On the Make a Map toolbar, right-click the Pick a State combo box and click View Source.

— Right-click here

You see the ThisDocument code module and wrapper lines for the combo box's SelectionChange event procedure. Code in SelectionChange runs when the user clicks a name in the drop-down list. (You also see code in the OpenDocument event that adds state names to the combo box with the AddItem method.)

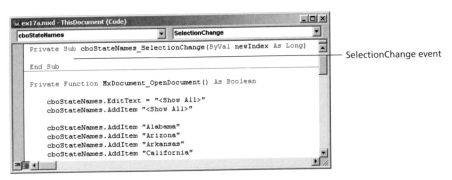

SelectionChange event

In the next step, you will declare and set a variable for the map document's IMxDocument interface. After that, you will set up IMap variables for the USA overview map and the EPA detail map.

5 In the SelectionChange event, declare and set a variable for the map document.

```
Dim pMxDoc As IMxDocument
Set pMxDoc = ThisDocument
```

The IMxDocument interface has the Maps property, which returns the IMaps interface of the collection of maps.

6 Declare and set a variable to get the map collection from IMxDocument.

```
Dim pMaps As IMaps
Set pMaps = pMxDoc.Maps
```

The IMaps interface has the Item property to get a map given its index position. The Item property returns IMap.

7 Declare and set two IMap variables for the EPA and USA maps.

```
Dim pEPAMap As IMap
Set pEPAMap = pMaps.Item(0)

Dim pUSAMap As IMap
Set pUSAMap = pMaps.Item(1)
```

```
ex17a.mxd - ThisDocument (Code)                          _ □ ×
cboStateNames              ▼    SelectionChange              ▼
   Private Sub cboStateNames_SelectionChange(ByVal newIndex As Long)
      Dim pMxDoc As IMxDocument
      Set pMxDoc = ThisDocument

      Dim pMaps As IMaps
      Set pMaps = pMxDoc.Maps

      Dim pEPAMap As IMap
      Set pEPAMap = pMaps.Item(0)

      Dim pUSAMap As IMap
      Set pUSAMap = pMaps.Item(1)
   End Sub
```

GETTING THE RIGHT MAP

Users can add, delete, or move maps and thereby change a map's index position number. So before setting the map variable, it's a good idea to confirm that you have the right map.

You can write a loop that tests each map in the collection to see which one it is. The IMaps interface doesn't have a Next method, like Enums do, so you can't use a Do While loop. However, since there is a specific number of maps to process, a For loop will work just as well.

You want to assign a unique variable to each map (for instance, a pEPAMap variable to refer to the EPA map and a pUSAMap variable to refer to the USA map). Since the Item property on IMaps returns IMap, and IMap has the Name property, you can test each map to see what its name is. You can then use the map's name to assign an appropriate variable.

```
Dim pMap As IMap
For X = 0 to 1
Set pMap = pMaps.Item(X)
     If pMap.Name = "EPA" Then
           Set pEPAMap = pMap
     End If
     If pMap.Name = "USA" Then
           Set pUSAMap = pMap
     End If
Next
```

With two distinct map variables, you can now get layers from either map. You will write code to get the Toxic sites layer and the Counties layer from the EPA map.

You get layers with the Layer property on IMap. (You did this in chapter 11, when you got layers to turn off their visibility.) The Layer property takes a layer's index position as an argument and returns the layer's ILayer interface.

Although ILayer is the returned interface, it doesn't have any properties or methods you want to use. You want IFeatureLayerDefinition for its DefinitionExpression property. Since the feature layers you are getting have access to both these interfaces, you can take the shortcut of letting VBA do QueryInterface for you. Instead of declaring variables to ILayer, you'll declare them directly to IFeatureLayerDefinition.

8 Declare and set IFeatureLayerDefinition variables for the Toxic sites layer and the Counties layer.

```
Dim pToxicLayerDef As IFeatureLayerDefinition
Set pToxicLayerDef = pEPAMap.Layer(0)

Dim pCountyLayerDef As IFeatureLayerDefinition
Set pCountyLayerDef = pEPAMap.Layer(1)
```

You now have a variable for each layer in the EPA map, pointing to the IFeatureLayerDefinition interface on FeatureLayer.

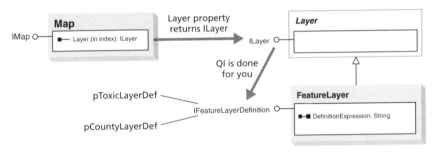

You want to filter out all toxic site and county features except those in the state picked by the user. The following lines create the query statement you need and assign it to a variable.

9 Declare and set a string variable to hold the query statement.

```
Dim strQuery As String
strQuery = "State_Name = '" & cboStateNames.EditText & "'"
```

```
ex17a.mxd - ThisDocument (Code)
cboStateNames                    SelectionChange

    Dim pUSAMap As IMap
    Set pUSAMap = pMaps.Item(1)

    Dim pToxicLayerDef As IFeatureLayerDefinition
    Set pToxicLayerDef = pEPAMap.Layer(0)

    Dim pCountyLayerDef As IFeatureLayerDefinition
    Set pCountyLayerDef = pEPAMap.Layer(1)

    Dim strQuery As String
    strQuery = "State_Name = '" & cboStateNames.EditText & "'"
End Sub
```

The user can display toxic sites and counties for all states by picking <Show All>. You will add an If Then statement that applies a definition query if a state is clicked and removes it if <Show All> is clicked.

10 Add the following If Then statement to determine if the user clicked <Show All>. Include the two comments.

```
If cboStateNames.EditText = "<Show All>" Then
'This code runs when the user clicks <Show All>


Else
'This code runs when the user clicks a state


End If
```

In this and the following exercises you will add code in the two commented areas.

11 Inside the If Then statement, after the first comment, set both layer definitions equal to an empty text string (empty double quotes with no space between them).

```
pToxicLayerDef.DefinitionExpression = ""
pCountyLayerDef.DefinitionExpression = ""
```

If the user clicks <Show All>, these two lines of code remove the definition queries from both layers, thereby displaying every state's toxic sites and counties.

12 After the Else keyword and the second comment, set each layer's DefinitionExpression property equal to the query string.

```
pToxicLayerDef.DefinitionExpression = strQuery
pCountyLayerDef.DefinitionExpression = strQuery
```

If the user clicks a state, these two lines of code set the definition query to show only the toxic states and counties for that state.

In the previous three chapters, every time you added a layer to a map, you refreshed the table of contents (pMxDoc.UpdateContents) and the map display (pMxDoc. ActiveView.Refresh). You haven't added a new layer here, so you don't need to do anything to the table of contents. You have changed the way layers draw, however, so you need to refresh the map display.

Instead of refreshing the layout's entire display area, you can get each map's IActiveView interface and refresh each map separately. The next two lines declare a variable to the EPA map's IActiveView and do QueryInterface from the pEPAMap variable that you declared to IMap in step 7.

13 Before the If Then statement, add the following lines of code to declare and set a variable for the EPA map's IActiveView interface.

```
Dim pEPAActiveView As IActiveView
Set pEPAActiveView = pEPAMap
```

In the next chapter, you will write code in the If Then statement that references this variable. That's why you need to put these lines before the If Then statement. The line of code that runs the Refresh method goes after the If Then statement because you want the display to redraw after the user picks a state or <Show All>.

14 After the If Then statement, use the Refresh method.

```
pEPAActiveView.Refresh
```

```
ex17a.mxd - ThisDocument (Code)
cboStateNames                          SelectionChange

    Dim strQuery As String
    strQuery = "State_Name = '" & cboStateNames.EditText & "'"

    Dim pEPAActiveView As IActiveView
    Set pEPAActiveView = pEPAMap

    If cboStateNames.EditText = "<Show All>" Then
    'This code runs when the user clicks <Show All>
        pToxicLayerDef.DefinitionExpression = ""
        pCountyLayerDef.DefinitionExpression = ""
    Else
    'This code runs when the user clicks a state
        pToxicLayerDef.DefinitionExpression = strQuery
        pCountyLayerDef.DefinitionExpression = strQuery
    End If

    pEPAActiveView.Refresh
End Sub
```

The code is ready to test.

15 Close Visual Basic Editor.

You will select a state from the combo box to test the SelectionChange event.

16 On the Make a map toolbar, click the Pick a State drop-down arrow and click Texas.

In the EPA map, only the counties and toxic sites in Texas draw. The view doesn't yet zoom in on Texas—you'll write that code in the next chapter.

When you set a feature layer's DefinitionExpression property, its query statement gets stored with the layer. You can see the query statement by opening the layer's Layer Properties window and looking at the Definition Query tab. The query for the Counties layer is shown below.

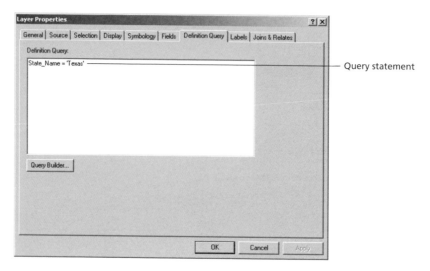

Now you will make sure the user can display the toxic sites and counties of all the states.

17 Click the Pick a State drop-down arrow, scroll to the top of the list, and click <Show All>.

After clicking <Show All>, all the counties and toxic sites draw.

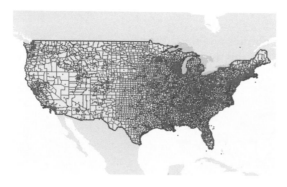

If you looked at the Counties layer's Definition Query tab now, it would have no query statement.

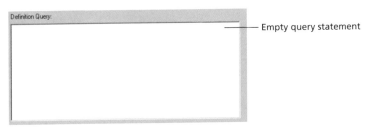

Empty query statement

18 If you want to save your work, click the File menu in ArcMap and click Save As. Navigate to **C:\ArcObjects\Chapter17**. Rename the file **my_ex17a.mxd** and click Save. If you are continuing with the next exercise, leave ArcMap open. Otherwise close it.

chapter

13
14
15
16
17
18
19
20

Selecting features and setting the selection color

In the last exercise, you wrote part of the code for the EPA detail map. In this exercise, you'll work on the USA overview map.

The overview map will highlight in red the U.S. state picked by the user. Since this map changes according to the same user selection as the EPA map, you'll be able to use the same strQuery variable you created in the previous exercise. Remember that you have also already set up a variable (pUSAMap) for working with the USA map.

You will write the code for making a feature selection. In a feature selection, all features draw, but the ones that are selected draw in a different color. To select and highlight features, you will use the SelectFeatures method on the IFeatureSelection interface. (As you'll see in chapter 18, this is not the only way to make a feature selection.)

The SelectFeatures method has three arguments: a query filter, a selection method, and the justOne argument.

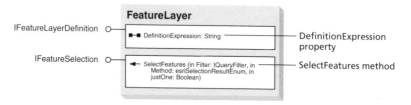

The first argument is a query filter, which is an object that you use to build and store query statements. (You can build queries without a query filter—you just did it in the last exercise—but query filters have useful properties when your queries become complex.) A query filter stores query statements in its WhereClause property.

You create a query filter with the code below.

```
Dim pFilter As IQueryFilter
Set pFilter = New QueryFilter
```

To select Arizona, you would set the query filter's WhereClause property as follows:

```
pFilter.WhereClause = "State_Name = 'Arizona'"
```

The second SelectFeatures argument is a selection method, which has five settings: esriSelectionResultNew, esriSelectionResultAdd, esriSelectionResultSubtract, esriSelectionResultAnd, and esriSelectionResultXOR.

Four of the five settings should be familiar, since they correspond to the interactive selection methods on the user interface.

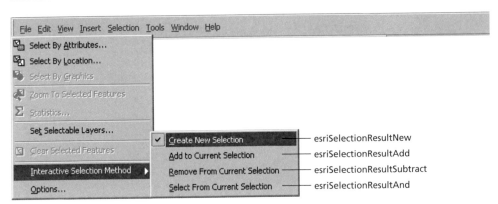

The fifth setting, esriSelectionResultXOR, performs an "exclusive or." This means that it reverses the current selection status of all features that satisfy the query.

The third argument is a Boolean (true/false) argument called justOne. When justOne is False, all features in the feature class are processed to see if they satisfy the query. When justOne is True, processing stops after the first feature that meets the query is found. The usual value is False, but True can be used when you know in advance that only one feature satisfies the query. (This saves unnecessary searching.) You can also use True if you simply want to confirm that a feature class is not empty of features that meet the query.

The code below shows the SelectFeatures method and its three arguments.

```
pFSLayer.SelectFeatures _
    pFilter, _
    esriSelectionResultNew, _
    True
```

Exercise 17b

In this exercise, you will continue developing the Toxic Sites application. You will write code to draw the selected U.S. state in red in the USA overview map.

1 Start ArcMap and open **ex17b.mxd** in the **C:\ArcObjects\Chapter17** folder.

You see the EPA and USA maps, and the Make a map toolbar. You will locate the Pick a State combo box's SelectionChange event procedure and add code to it.

2 Open the Customize dialog box.

3 On the Make a map toolbar, right-click the Pick a State combo box and click View Source.

You see the ThisDocument code module and the combo box's SelectionChange event procedure.

4 In the SelectionChange event procedure, scroll down until you locate the If Then statement.

You will add code before, inside, and after the If Then statement. Before the statement, your code will get the States layer from the USA map. Inside the statement, you'll clear the user-selected state when the user clicks <Show All> or highlight the user-selected state in red. After the statement, you'll refresh the USA map.

To get the States layer, you'll use IMap's Layer property. (Your pUSAMap variable from the previous exercise already points to this interface.) As before, you don't have any need for the ILayer interface that the Layer property returns. This means that you can again skip over this interface by letting VBA do QueryInterface for you. In the last exercise, you took the shortcut by declaring your layer variables directly to IFeatureLayerDefinition. This time, you'll declare a layer variable to IFeatureSelection, which has the SelectFeatures and Clear methods you need.

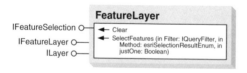

5 Immediately before the If Then statement, declare an IFeatureSelection variable and set it equal to the States layer in the USA map.

```
Dim pUSALayer As IFeatureSelection
Set pUSALayer = pUSAMap.Layer(0)
```

Next you will write code in both parts of the If Then statement. You will start by writing code that clears the selected state when the user clicks <Show All>.

6 Scroll down in the If Then statement and locate the line of code that clears the definition expression for the Counties layer by setting it equal to empty quotes. (It's the last line in the first part of the If Then statement.)

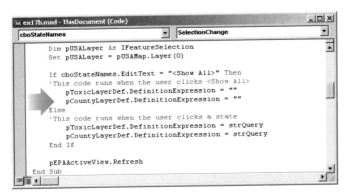

You will add code after this line to clear the selection in the overview map when the user clicks <Show All>. The code that is there from the previous exercise clears the definition queries on the EPA map. The line you are adding now keeps the USA and EPA maps in synch.

7 Immediately before the Else keyword, add the following line to clear any selected state in the USA map's States layer.

```
pUSALayer.Clear
```

Next you will add code to the other part of the If Then statement to select (highlight) the state picked by the user and set its color to red.

8 In the If Then statement, locate the line of code that sets the definition query for the Counties layer. (It's the last line in the second part of the If Then statement.)

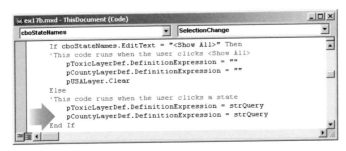

The code from the previous exercise sets the definition queries for the EPA map. Again, your new code will keep the maps in synch.

The SelectFeatures method takes a query filter as its first argument, so you need to create a query filter. You'll set the filter's WhereClause property equal to the query statement you wrote in the last exercise.

9 Immediately before the End If keywords, create a query filter.

```
Dim pFilter As IQueryFilter
Set pFilter = New QueryFilter
```

10 Add a line of code to set the query filter's WhereClause equal to the query string.

```
pFilter.WhereClause = strQuery
```

Now you can run the SelectFeatures method on the States layer.

11 Make the selection with the SelectFeatures method.

```
pUSALayer.SelectFeatures _
    pFilter, esriSelectionResultNew, True
```

The first argument is the query filter. The second argument is esriSelectionResultNew because you want a brand new selection each time the user picks a different state. The third argument is set to True because only one feature meets the query. (There is no need to search for more than one state called New Mexico, for example.) Setting this argument to False would give the same result—the search would just take a little longer.

Selected features draw in the current selection color (by default, a cyan outline.) You will set the SelectionColor property, shown below on IFeatureSelection, to red.

12 Create an RgbColor and set its Red property to 200.

```
Dim pRedColor As IRgbColor
Set pRedColor = New RgbColor
pRedColor.Red = 200
```

13 Set the States layer's SelectionColor property equal to the RgbColor object.

```
Set pUSALayer.SelectionColor = pRedColor
```

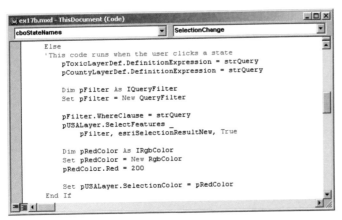

To see your drawing instructions take effect, you have to refresh the USA map's display.

Your pUSAMap variable still points to IMap, so you will use QueryInterface to get its IActiveView interface.

14 Immediately after the If Then statement, add the following code to get the USA map's IActiveView interface and run its Refresh method.

```
Dim pUSAActiveView As IActiveView
Set pUSAActiveView = pUSAMap

pUSAActiveView.Refresh
```

In the previous exercise, you declared the pEPAActiveView variable before the If Then statement and ran its Refresh method after the statement. You did that because, in the next chapter, you will use pEPAActiveView inside the If Then statement. That's not the case with pUSAActiveView, so you can keep these three lines together and all after the If Then statement.

15 Close Visual Basic Editor.

You will test the code by picking a state in the combo box.

16 On the Make a map toolbar, click the Pick a State drop-down arrow and click New Mexico.

In the overview map, New Mexico draws in red. In the detail map, New Mexico's toxic sites and counties are displayed. In the next chapter, you will add code to make the detail map zoom in on the selected state.

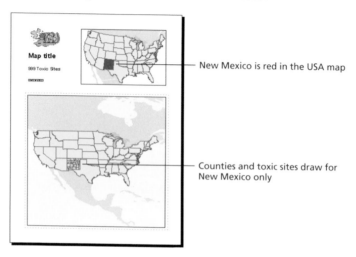

New Mexico is red in the USA map

Counties and toxic sites draw for New Mexico only

Now you'll select another state to make sure that both maps update and refresh correctly.

17 Click the Pick a State drop-down arrow and click Wyoming.

Wyoming highlights in the overview map and displays its counties and toxic sites in the detail map. Next, you will click <Show All> to make sure that the USA map clears the feature selection.

18 Click the Pick a State drop-down arrow, scroll all the way to the top, and click <Show All>.

In the USA map, no states are highlighted. In the EPA map, all counties and toxic sites display.

19 If you want to save your work, click the File menu in ArcMap and click Save As. Navigate to **C:\ArcObjects\Chapter17**. Rename the file **my_ex17b.mxd** and click Save. If you are continuing with the next chapter, leave ArcMap open. Otherwise close it.

Working with selected features

In the last chapter, you wrote code to select features, but you didn't do anything with these features except draw them. In this chapter, you will go further and work with selected features both as a group and individually.

To work with selected features as a group, you get or make a selection set. A selection set is not too glamorous—it's just a container that you can put selected features in or take them out of. Unlike other collection objects you've worked with (Enums and map collections), a selection set doesn't have a method for getting individual objects from the collection. In fact, there isn't a whole lot you can do with it, but it does have one important property: it can give you a count of the features it contains. You'll use this Count property to report the number of toxic sites for the U.S. state picked by the user.

To work with selected features individually, you make a cursor. A cursor is like an Enum, with a pointer and a method to move from one object to the next. (A cursor could have been called EnumSelectedFeatures, or something like that, but "cursor" is a standard term in the database world for this kind of an object collection.)

Cursors give you access to a feature's spatial and attribute information. Say you get the Arizona feature from a cursor of selected U.S. states. You could go on to find out things like its area, perimeter, centroid, and the x,y coordinates that make up its polygon vertices. Or you could find out its population, per capita income, or number of mobile homes—anything that is stored in the attribute table. You can also set the spatial and attribute information. You might replace an old population figure with a new one, for example. You'll get a chance to edit data values in chapter 20.

You might think of selection sets as Clark Kent and cursors as Superman—one is weak and the other is powerful, and you can make a cursor from a selection set. You can also make a cursor independently of a selection set, however, which is the way you'll do it in this chapter.

Selection sets and cursors are made up of records. That may sound a bit odd, considering that we've been talking about features, but the term "record" actually refers both to rows in a table and to features in a feature class. That's because, as the following diagram shows, a feature class is really a type of table and a feature is a type of row. The diagram also shows that feature classes are composed of features just as tables are composed of rows.

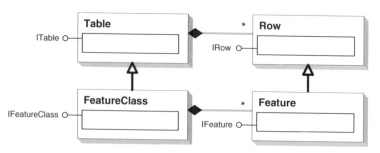

From the user's point of view, a feature class can be displayed either as features on a map or as records in a table. A selected feature can be displayed as a highlighted piece of geometry or as a highlighted row in a table.

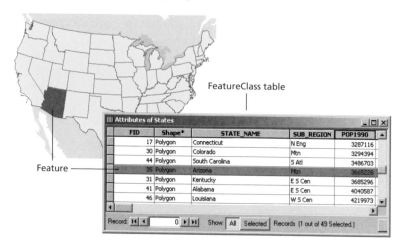

In this chapter, you will continue building the toxic sites application. In the first exercise, you will use a selection set to get a count of toxic sites in the U.S. state picked by the user. In the second exercise, you will use a cursor to get information about the selected state's spatial extent. You'll use that information to zoom the detail map in on the selected state.

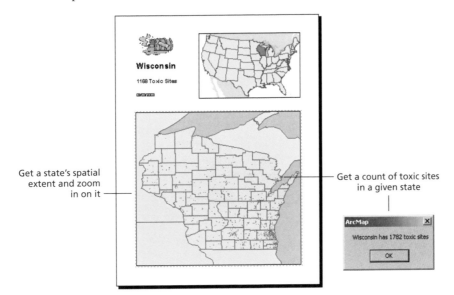

Get a state's spatial extent and zoom in on it

Get a count of toxic sites in a given state

Using selection sets

Every feature layer has a selection set. It may contain a single feature, many features, or every feature in the layer. If no features are selected, the selection set is still there, but it is empty.

When a user clicks on features with the Select Features tool or uses the Selection menu, they are defining a selection set. Programmers define a selection set when they write code to select features, as you did in the last chapter. No matter how the selection set is built, you can get it with the SelectionSet property on FeatureLayer's IFeatureSelection interface.

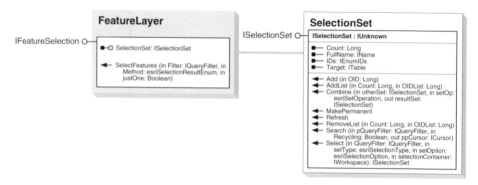

Say you wanted to get the selection set shown in the following graphic:

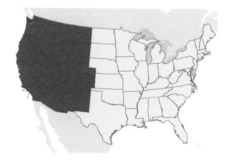

In the last chapter, you used IFeatureSelection for its SelectFeatures method, so the code to get this interface should look familiar. You get the layer by using the Layer property on IMap. Then you declare a variable to IFeatureSelection and let VBA do QueryInterface from ILayer.

```
Dim pMxDoc As IMxDocument
Set pMxDoc = ThisDocument
Dim pMap As IMap
Set pMap = pMxDoc.FocusMap
Dim pFLayer As IFeatureSelection
Set pFLayer = pMap.Layer(0)
```

To get the selection set, you then declare a variable to ISelectionSet and set it equal to the SelectionSet property.

```
Dim pWestSelectionSet As ISelectionSet
Set pWestSelectionSet = pFLayer.SelectionSet
```

A layer is not limited to a single selection set. It can have as many as you want to make variables for. Say a user working with the U.S. states layer decides to make a new feature selection.

You could get this selection set, too, and store it in a second variable.

```
Dim pEastSelectionSet As ISelectionSet
Set pEastSelectionSet = pFLayer.SelectionSet
```

Although a feature layer can have more than one selection set, only one can be displayed at a time. To switch back and forth between selection sets, you set the feature layer's SelectionSet property. This is a byRef property (open barbell) that requires the Set keyword.

```
Set pFLayer.SelectionSet = pWestSelectionSet
```

To see the new selection set display, you have to refresh the map's active view.

```
pMxDoc.ActiveView.Refresh
```

In the example above, it was assumed that a user was making the selections on the user interface. You can also create a selection set directly with a query filter and a table, as shown in the following diagram.

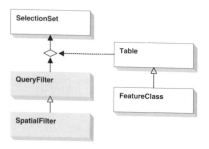

This is the first time you have seen the open diamond symbol. It means that two objects are needed to create a third. Dashed arrows point from Table and QueryFilter to the diamond, and from the diamond to SelectionSet. This means that to create a selection set, you need both a query filter and a table (or a feature class, since it is a type of table).

When you have these objects, you make the selection set by running the Select method on IFeatureClass, as shown below. The Select method has four arguments and returns the ISelectionSet interface of a selection set.

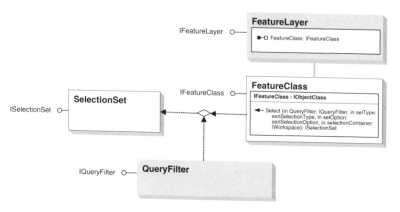

The first argument is a query filter, which you know about from the last chapter.

The second argument is a selection type, which has three choices. The first choice, esriSelectionTypeIDset, specifies that the ID numbers of features in the selection set be written to a database table, such as an SDE® logfile. The second choice, esriSelectionTypeSnapshot, specifies that these IDs be held in computer memory instead. (The first choice is better for large selection sets and the second for small ones.) The third choice, esriSelectionTypeHybrid, makes the decision for you based on the size of the selection set.

The third argument is a selection option, which also has three choices: esriSelectionOptionNormal, esriSelectionOptionOnlyOne, and esriSelection-OptionEmpty. This argument is similar to the justOne argument that you used with the SelectFeatures method in the last chapter. The Normal choice selects all features that meet the query filter's criteria.

The fourth argument specifies a workspace for storing the table created by the second argument. The value Nothing puts the table into the same workspace as the feature class. (This argument is required even if you use esriSelectionTypeSnapshot as the selection type.)

Exercise 18a

In this exercise, you will write code to create a selection set and get a count of all toxic sites in the state picked by the user. For now, you'll report the count in a message box. In the next chapter, you will integrate the count as text graphics on a map.

1 Start ArcMap and open **ex18a.mxd** in the **C:\ArcObjects\Chapter18** folder.

You see the USA and EPA maps on the layout and the Make a map toolbar.

2 Open the Customize dialog box. On the Make a map toolbar, right-click the Pick a State combo box and click View Source.

You see the ThisDocument code module and the SelectionChange event procedure.

3 In the SelectionChange event, scroll down to the last line of the If Then statement (the one that sets the selection color of the USA map's States layer).

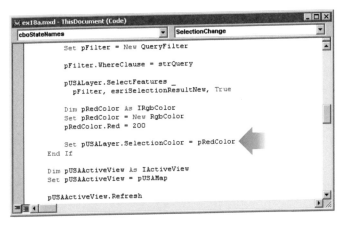

```
ex18a.mxd - ThisDocument (Code)

cboStateNames                          SelectionChange

        Set pFilter = New QueryFilter

        pFilter.WhereClause = strQuery

        pUSALayer.SelectFeatures _
            pFilter, esriSelectionResultNew, True

        Dim pRedColor As IRgbColor
        Set pRedColor = New RgbColor
        pRedColor.Red = 200

        Set pUSALayer.SelectionColor = pRedColor
    End If

    Dim pUSAActiveView As IActiveView
    Set pUSAActiveView = pUSAMap

    pUSAActiveView.Refresh
```

You will add code after this line to create a selection set.

A selection set requires a query filter and a table. Part of the work is already done, because you can use the query filter (pFilter) that you made in the last chapter. Its query statement finds features whose State_Name attribute matches the state picked in the combo box. When you apply pFilter to the Toxic sites layer, it will find all toxic sites within the selected state.

The other component you need is the feature class of the Toxic sites layer. If you scroll up toward the top of the SelectionChange event procedure, you'll see that you already have a variable for this layer, pToxicLayerDef, which you set in the last chapter.

pToxicLayerDef is declared to IFeatureLayerDefinition, but to get the layer's feature class, you need IFeatureLayer. So you will switch interfaces.

You need a variable to point here to get to the FeatureClass property ——— IFeatureLayer ○—

pToxicLayerDef ——— IFeatureLayerDefinition ○—

FeatureLayer

■□ FeatureClass: IFeatureClass

4 After the line above that sets the selection color to red, and immediately before the End If keywords, declare an IFeatureLayer variable and set it equal to pToxicLayerDef.

```
Dim pToxicFLayer As IFeatureLayer
Set pToxicFLayer = pToxicLayerDef
```

Now you can get the feature class of the Toxic sites layer, using the FeatureClass property on IFeatureLayer.

5 Declare and set an IFeatureClass variable.

```
Dim pToxicFClass As IFeatureClass
Set pToxicFClass = pToxicFLayer.FeatureClass
```

You have the two objects you need. To make the selection set, you'll run the Select method on the feature class. For the method's first argument, you'll use pFilter (your query filter). For the second argument, you'll use the hybrid choice and let VBA decide how to store the IDs of the selected features. For the third argument, you'll choose the normal option, which selects all features that satisfy the query (all toxic sites within the U.S. state the user picks). You'll set the fourth argument to Nothing.

6 Declare and set an ISelectionSet variable.

```
Dim pSelectionSet As ISelectionSet
Set pSelectionSet = pToxicFClass.Select _
    (pFilter, _
    esriSelectionTypeHybrid, _
    esriSelectionOptionNormal, _
    Nothing)
```

You will use ISelectionSet's Count property to find out how many toxic sites are in the selection set. You'll report the count in a message box.

7 Use a message box to display the state name and the number of toxic sites it contains.

```
MsgBox cboStateNames.EditText & " has " _
    & pSelectionSet.Count & " toxic sites"
```

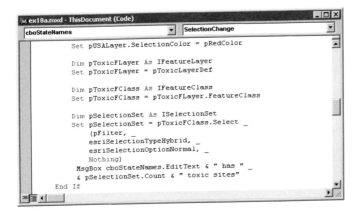

```
Set pUSALayer.SelectionColor = pRedColor

Dim pToxicFLayer As IFeatureLayer
Set pToxicFLayer = pToxicLayerDef

Dim pToxicFClass As IFeatureClass
Set pToxicFClass = pToxicFLayer.FeatureClass

Dim pSelectionSet As ISelectionSet
Set pSelectionSet = pToxicFClass.Select _
    (pFilter, _
    esriSelectionTypeHybrid, _
    esriSelectionOptionNormal, _
    Nothing)
MsgBox cboStateNames.EditText & " has " _
    & pSelectionSet.Count & " toxic sites"
End If
```

The code is ready to test.

8 Close Visual Basic Editor.

9 Click the Pick a State drop-down arrow and click California.

In the next chapter, you will write code to put this number in the map's title area.

10 Click OK.

The message box closes. California is highlighted in the overview map and its counties and toxic sites display in the detail map.

Although the California toxic sites are contained in this selection set, they do not highlight on the EPA map. Creating (or getting) a selection set does not highlight features. If you want the features to draw in the selection color, you could set the SelectionSet property on IFeatureSelection. That would mean declaring a variable to IFeatureSelection and setting it equal to your pToxicFLayer variable, then setting the SelectionSet property equal to your pSelectionSet variable.

```
Dim pToxicFeatSel As IFeatureSelection
Set pToxicFeatSel = pToxicFLayer
Set pToxicFeatSel.SelectionSet = pSelectionSet
```

11 If you want to save your work, click the File menu in ArcMap and click Save As. Navigate to **C:\ArcObjects\Chapter18**. Rename the file **my_ex18a.mxd** and click Save. If you are continuing with the next exercise, leave ArcMap open. Otherwise close it.

Using cursors

To work with individual records or features—to get or set their attribute and spatial information—you create a cursor. A cursor is a group of records organized in rows. It's like a table that you would open in ArcMap, except that you don't actually see it.

A cursor, like a selection set, is created from a query filter and a table. A feature cursor is a special type of cursor that you use with feature classes.

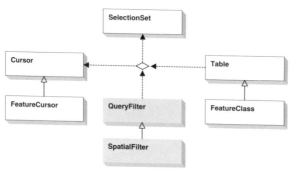

In the last exercise, you ran the Select method on IFeatureClass to make a selection set. The same interface has various methods to make a feature cursor.

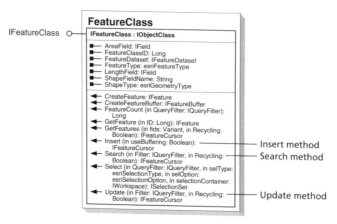

The Insert and Update methods let you add new features to a feature class or edit existing features. You'll use the Update method in chapter 20. In this exercise, you will use the Search method, which makes a cursor that contains all features satisfying a query statement. You use this method when you want to get the attribute or spatial information of particular features but don't intend to make new features. Although Insert, Update, and Search are the names of methods that create cursors, the cursors themselves are often called by these names. For instance, a cursor created by the

Insert method can be called an insert cursor. Given a feature class, pStatesFClass, and a query filter, pFilter, you create a search cursor with the two lines of code below.

```
Dim pFCursor As IFeatureCursor
Set pFCursor = pStatesFClass.Search(pFilter, True)
```

Once you have created a cursor, you work with it in the same way you work with an Enum (like the one in chapter 15). The cursor has a pointer that starts off pointing at the top of the collection of features.

Pointer ────────▶ Pointing to nothing at the top of the cursor

STATE_NAME	STATE_FIPS	POP1997	POP90_SQMI
Alabama	01	4298715	78
Arizona	04	4528866	32
Arkansas	05	2529864	44
California	06	32197302	189
Colorado	08	3885615	32
Connecticut	09	3277113	661
Delaware	10	731218	324

Cursor object contains a collection of records from a table

There is a NextFeature method, on the IFeatureCursor interface, that lets you move through the features.

IFeatureCursor O—

FeatureCursor
◄— NextFeature: IFeature

NextFeature method

The following code takes you from the starting point to the first feature in the cursor:

```
Dim pFeature As IFeature
Set pFeature = pFCursor.NextFeature
```

pFeature ——

STATE_NAME	STATE_FIPS	POP1997	POP90_SQMI
Alabama	01	4298715	78
Arizona	04	4528866	32
Arkansas	05	2529864	44
California	06	32197302	189
Colorado	08	3885615	32
Connecticut	09	3277113	661
Delaware	10	731218	324

The NextFeature method returns a feature's IFeature interface. As the following diagram shows, this interface gives you access to the spatial properties of a feature. One of these is the Extent property, which returns a feature's envelope. Every feature has an extent envelope that represents the smallest possible rectangle that can surround the feature.

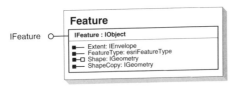

IFeature O—

Feature
IFeature : IObject

■— Extent: IEnvelope
■— FeatureType: esriFeatureType
■—◻ Shape: IGeometry
■— ShapeCopy: IGeometry

The graphic below illustrates the envelope for a polygon feature of Wisconsin.

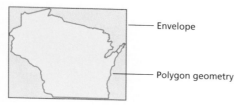

— Envelope

— Polygon geometry

Say you are pointing at this feature (pFeature) with a cursor. The next lines of code get its envelope.

```
Dim pEnvelope As IEnvelope
Set pEnvelope = pFeature.Extent
```

Features have envelopes and so do layers and the active view. Zooming in and out on a map depends on envelopes. The current zoom setting is determined by the active view's extent envelope illustrated below.

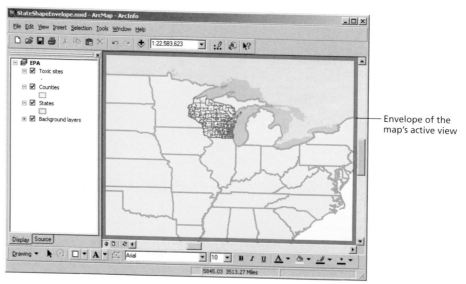

— Envelope of the map's active view

When you want to zoom in or out, you set the active view's envelope equal to some new envelope, such as the Wisconsin feature's envelope (pEnvelope).

You get or set the active view's envelope with the Extent property on IActiveView.

IActiveView O—

Map

■—■ Extent: IEnvelope ——— Extent property

As long as you already have a variable, pMap, that points to IMap, you can do QueryInterface to switch to the map's IActiveView interface.

```
Dim pMapsActiveView As IActiveView
Set pMapsActiveView = pMap
```

You can then set the active view's envelope.

```
pMapsActiveView.Extent = pEnvelope
```

After you refresh the view, you will be zoomed in on Wisconsin.

The state's envelope is equal to the active view's envelope

Exercise 18b

In this exercise, you will write more code for the EPA detail map. You will make a feature cursor that contains the U.S. state chosen by the user. You will get this state's envelope and set the envelope of the map's active view equal to it. The result will be that the detail map zooms in on the state that the user picks. When the user picks <Show All>, you will zoom back out to the extent of the States layer.

1 Start ArcMap and open **ex18b.mxd** in the **C:\ArcObjects\Chapter18** folder.

You see the USA and EPA maps on the layout and the Make a map toolbar.

2 Open the Customize dialog box. On the Make a map toolbar, right-click the Pick a State combo box and click View Source.

You see the ThisDocument code module and the SelectionChange event procedure.

You will write the code that zooms out to the extent of the States layer first. To do that, you need a variable that refers to the States layer in the EPA map. (You have variables for the Counties and Toxic sites layers, but not the States layer.)

3 In the SelectionChange event procedure, find the line of code that sets the pCountyLayerDef variable for the Counties layer in the EPA map. (It's about the twelfth line of code in the procedure.)

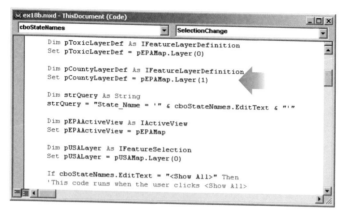

You will add code after this line.

4 Declare and set a variable for the States layer using the ILayer interface.

```
Dim pStatesLayer As ILayer
Set pStatesLayer = pEPAMap.Layer(2)
```

Since the States layer is the third layer on the EPA map, you use position 2 as the Layer property's argument.

Now you will add code to the If Then statement. When the user clicks <Show All>, you will get the States layer's envelope and use it to set the extent of the active view.

Features and the active view have an Extent property to return their envelopes. Layers have a property called AreaOfInterest that does the same thing.

5 Inside the If Then statement, locate the line of code that clears the selection in the USA Layer when the user clicks <Show All>.

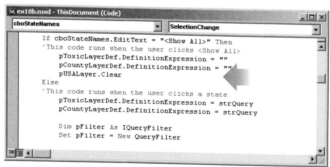

Your code will go after this line.

You don't need to declare a variable to the map's IActiveView, because you already have one (pEPAActiveView). You set it up in the previous chapter to refresh the view when the user picked a new state. You may recall that you were going to reference this variable from inside the If Then statement in this chapter. This is where you do it. Inside the two parts of the If Then statement, you will change the active view's extent envelope, causing it to zoom in or out.

6 Immediately after the Clear method and before the Else keyword, set the map's active view extent equal to the area of interest of the States layer.

```
pEPAActiveView.Extent = pStatesLayer.AreaOfInterest
```

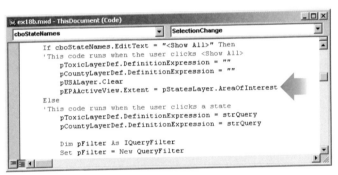

Now you'll add the code for zooming in on a U.S. state to the second part of the If Then statement.

7 Scroll down to the message box lines of code at the end of the If Then statement.

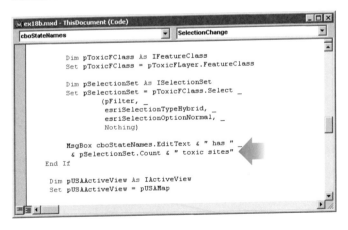

Your code will go after this line.

To create a cursor, you need a query filter and a table (feature class). These are the same objects you needed for a selection set in the previous exercise. Once again, you will use the existing query filter and, once again, you will switch interfaces, this time from ILayer to IFeatureLayer to get the layer's feature class.

8 Immediately after the MsgBox code and before the End If keywords, declare and set an IFeatureLayer variable for the States layer.

```
Dim pStatesFLayer As IFeatureLayer
Set pStatesFLayer = pStatesLayer
```

Now you can get the feature class of the States layer.

9 Declare and set a variable for the States layer's feature class.

```
Dim pStatesFClass As IFeatureClass
Set pStatesFClass = pStatesFLayer.FeatureClass
```

You have both the objects you need and can run the Search method to create a feature cursor.

10 Declare and set a variable to create a feature cursor with the FeatureClass and QueryFilter objects.

```
Dim pFCursor As IFeatureCursor
Set pFCursor = pStatesFClass.Search(pFilter, True)
```

The Search method has two arguments. The first is the query filter. The second, called recycling, specifies the way in which features in a cursor are held in memory during processing. You will learn about the recycling argument in chapter 20; for now, just set it to true.

The cursor you have made contains only a single feature. (In the States layer, there is only one feature corresponding to each State_Name attribute value.) Even so, you still deal with the cursor as a collection. Therefore, you will move the pointer with the NextFeature method, which returns IFeature.

11 Declare an IFeature variable for the selected U.S. state and use the NextFeature method to set it.

```
Dim pFeature As IFeature
Set pFeature = pFCursor.NextFeature
```

Now you are pointing to the feature object in the cursor. You want to get its envelope and set the extent of the map's active view equal to it.

12 Declare and set a variable to hold the selected U.S. state's envelope.

```
Dim pEnvelope As IEnvelope
Set pEnvelope = pFeature.Extent
```

If you use this envelope as is to set the active view's extent, the corners of the state might touch the sides of the map, as in the following graphic:

State boundary touches edge of the active view

You will put a little space between the feature envelope and the edge of the map. One way to do this would be to run the Execute method on the Fixed Zoom Out command. (You learned this technique in chapter 13.)

Another way is to use IEnvelope's Expand method to expand the envelope. The Expand method has three arguments. The first two are width and height settings. The first value expands the envelope in the x direction, the second in the y direction. (The center point does not change.) The third argument interprets the two values as ratios or as map units. It is a Boolean argument and you will use True to get ratios.

13 Expand the envelope to 1.1 times its original size.

```
pEnvelope.Expand 1.1, 1.1, True
```

14 Set the extent of the map's active view equal to the expanded envelope.

```
pEPAActiveView.Extent = pEnvelope
```

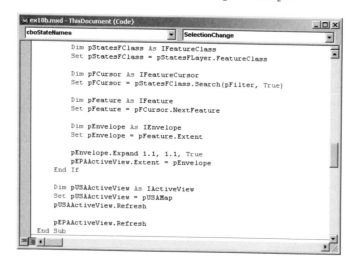

The code to refresh the EPA map is already there, after the If Then statement. You wrote this code in the previous chapter.

The code is ready to test.

15 Close Visual Basic Editor.

16 On the Make a map toolbar, click the Pick a State drop-down arrow and click Wisconsin.

17 Click OK.

The EPA map zooms in on Wisconsin and you see its counties and toxic sites. You also see that it is zoomed a little bit to leave space between Wisconsin's polygon boundary and the map edge.

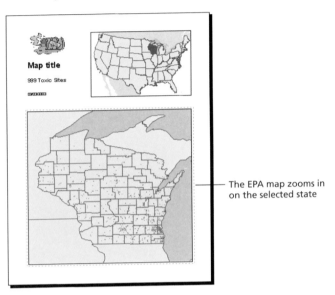

The EPA map zooms in on the selected state

18 Click the Pick a State drop-down arrow and click <Show All>.

The EPA map zooms out and you see all counties and all toxic sites. In the following chapter, you will finish your work on the application by coding the map title and other elements to update according to the user's selection.

19 If you want to save your work, click the File menu in ArcMap and click Save As. Navigate to **C:\ArcObjects\Chapter18**. Rename the file **my_ex18b.mxd** and click Save. If you are continuing with the next chapter, leave ArcMap open. Otherwise close it.

Making dynamic layouts

Naming elements
Manipulating text elements

Everything you add to a layout—a data frame, a scale bar, a north arrow, text, a picture, a neatline, and so on—is an element. In other words, a layout is composed of elements. The abstract Element class has FrameElement and GraphicElement below it. (You may recall this from chapter 12, when you worked with marker elements, a type of graphic element.)

Frame elements include both data frames and elements related to them. Data frames are represented by the MapFrame subclass. Associated with a MapFrame are its MapSurroundFrames, which are elements that change along with the data frame. When the data frame below is rotated and zoomed to Italy, its north arrow and scale bar (map surround frames) change to reflect the new orientation and scale.

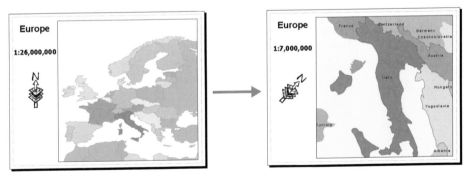

The map title, however, does not change, even though the map is now a map of Italy rather than Europe. Graphic elements, such as text, graphics, pictures, and fills, are not associated with map frames or other layout elements. Once set, they don't change unless they are told to.

Your job in this chapter is to make graphic elements (specifically, text elements) change in response to what is being displayed in a map frame. When the user picks a U.S. state in the combo box, your code will display the state name as the map title and the number of its toxic sites as the subtitle.

The following diagram shows the relationship of elements on the layout page. FrameElement has several subclasses, but only MapFrame and MapSurroundFrame are shown. The line connecting MapFrame and MapSurroundFrame tells you that they are associated. (That's why when you rotate a map frame, for example, its north arrow rotates with it.)

GraphicElement also has many subclasses, but TextElement is the only one you'll use in this chapter.

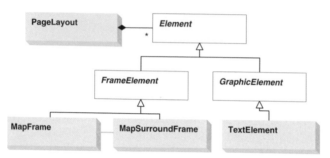

When you work with layout elements, some of the things you do are specific to the properties and methods of the element. For example, you might get a line element's Symbol property to change its color, or a marker element's Symbol property to change its size. But you also work with elements in a more general way by selecting and unselecting them, adding and deleting them, bringing them forward, sending them back, and so on.

The PageLayout coclass (below) has two interfaces for working with elements at this general level: IGraphicsContainer and IGraphicsContainerSelect. In spite of their names, these interfaces apply to all layout elements, frame elements and graphic elements alike.

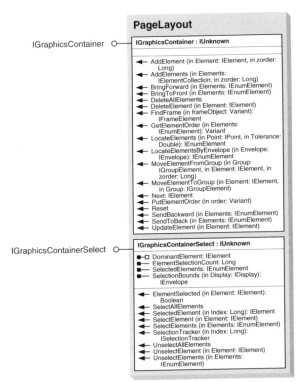

PageLayout

IGraphicsContainer ○——

IGraphicsContainer : IUnknown

- ◄— AddElement (in Element: IElement, in zorder: Long)
- ◄— AddElements (in Elements: IElementCollection, in zorder: Long)
- ◄— BringForward (in Elements: IEnumElement)
- ◄— BringToFront (in Elements: IEnumElement)
- ◄— DeleteAllElements
- ◄— DeleteElement (in Element: IElement)
- ◄— FindFrame (in frameObject: Variant): IFrameElement
- ◄— GetElementOrder (in Elements: IEnumElement): Variant
- ◄— LocateElements (in Point: IPoint, in Tolerance: Double): IEnumElement
- ◄— LocateElementsByEnvelope (in Envelope: IEnvelope): IEnumElement
- ◄— MoveElementFromGroup (in Group: IGroupElement, in Element: IElement, in zorder: Long)
- ◄— MoveElementToGroup (in Element: IElement, in Group: IGroupElement)
- ◄— Next: IElement
- ◄— PutElementOrder (in order: Variant)
- ◄— Reset
- ◄— SendBackward (in Elements: IEnumElement)
- ◄— SendToBack (in Elements: IEnumElement)
- ◄— UpdateElement (in Element: IElement)

IGraphicsContainerSelect ○——

IGraphicsContainerSelect : IUnknown

- ■—□ DominantElement: IElement
- ■— ElementSelectionCount: Long
- ■— SelectedElements: IEnumElement
- ■— SelectionBounds (in Display: IDisplay): IEnvelope

- ◄— ElementSelected (in Element: IElement): Boolean
- ◄— SelectAllElements
- ◄— SelectedElement (in Index: Long): IElement
- ◄— SelectElement (in Element: IElement)
- ◄— SelectElements (in Elements: IEnumElement)
- ◄— SelectionTracker (in Index: Long): ISelectionTracker
- ◄— UnselectAllElements
- ◄— UnselectElement (in Element: IElement)
- ◄— UnselectElements (in Elements: IEnumElement)

IGraphicsContainer has methods for adding, deleting, and reordering elements. Like the Enums and cursors that you have used before, it also has a Next method for moving through a collection of elements one at a time. (Although IGraphicsContainer is an interface, not an object, it is commonly called "the graphics container" as if it were an object.)

IGraphicsContainerSelect has a DominantElement property for getting the selected element. It also has an ElementSelectionCount property that tells you how many elements are selected.

In this chapter, you'll use both these interfaces to help you get and update the text elements on the map.

Naming elements

In this chapter, you will update the text elements shown below based on the code you've written in the previous two chapters.

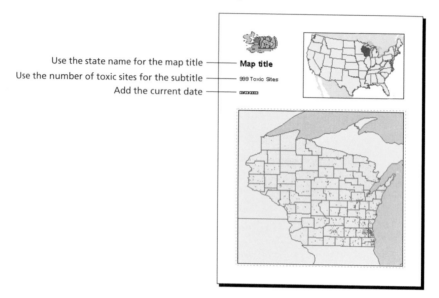

Use the state name for the map title — **Map title**
Use the number of toxic sites for the subtitle — 999 Toxic Sites
Add the current date —

It sounds pretty easy. You have a layout full of different elements and you have to get a few of them and update them. That should be a simple matter of picking the right elements out of the layout's graphics container, and setting the relevant properties equal to your existing variables for the state name and toxic site counts.

The complication is that layout elements don't have unique identifiers built in to them. So although it's easy for you to tell one element from another when you look at the user interface, it's not easy for VBA. You can't get an element by an index position number, like you can with maps or layers. And elements don't have pre-defined names the way style gallery items do.

One idea that might occur to you is writing some TypeOf statements, like you did in chapter 12, to see what interfaces an element has. That could work to tell different types of elements apart—map frame elements from text elements, for example. But in this situation, the elements you need to update (title, subtitle, and date) are all text elements. A Type Of statement won't help you there.

Fortunately, all layout elements have an interface called IElementProperties2, on the Element abstract class, that can be of service. This interface has a Name property you can set and get. Setting an element's name gives you a unique characteristic by which to get it from the graphics container.

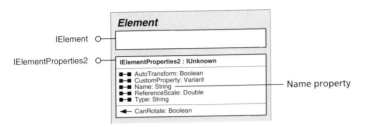

A handy way to employ this property could be with a UIButton. After interactively selecting the element you want on the layout, you could click the button to open an input box and set the element's name. It would also be nice to have a GetName button, so you could check and see whether or not you've already named an element.

Making these two buttons is what you're going to do in this exercise. For the SetName button, you'll use the DominantElement property on IGraphicsContainerSelect to get whichever element is selected on the layout. You'll make an input box to type in a name for this element, then you'll set the Name property on IElementProperties2. You'll enclose your code in an If Then statement that makes sure only one element is selected at a time.

Your code for the GetName button will be very similar. You'll get the selected element, get its name, and report the name in a message box.

Besides working with a couple of new interfaces and objects, this exercise gives you a chance to practice some things you did earlier in the book, like making UIButtons and using input boxes. For the first time, you'll be saving your work to the normal template, rather than to an .mxd file.

Exercise 19a

By assigning names to elements, you make it possible to get the ones you want from the graphics container. Since ArcMap doesn't have buttons to set or get an element's name, you'll make these buttons yourself. Then you'll use them to set the names of the title, subtitle, and date elements on the toxic sites map.

1 Start ArcMap and open **ex19a.mxd** in the **C:\ArcObjects\Chapter19** folder.

When the map opens, you see the toxic sites map from the previous chapters.

2 Open the Customize dialog box. Click the Commands tab. In the Categories list, click UIControls.

The buttons you're about to create can help you in other projects when you need to get elements from a layout. You'll save the customizations to your normal template (Normal.mxt), so you'll have these buttons on your layout toolbar no matter which map document you open.

Not saving the buttons to the .mxd file also means that the user won't see them. (Your normal template is specific to your installation of ArcMap, so the user has a different one.) That's good—the user doesn't have any need for these buttons.

3 For Save in, select Normal.mxt.

Now you'll create the first button.

4 Click the New UIControl button and click Create.

And now the second button.

5 Click the New UIControl button again and click Create.

In the Commands list, you see Normal.UIButtonControl1 and Normal.UIButtonControl2.

6 In the Commands list, rename the two new buttons **Normal.SetElementName** and **Normal.GetElementName**.

7 Drag both buttons onto the right side of the Layout toolbar.

SetElementName   GetElementName

You will put text on each button and add line separators.

8 On the Layout toolbar, right-click the first button and click Text Only. Right-click it again and click Begin a Group. Do the same for the second button.

Now you will write code for the two buttons.

9 Right-click SetElementName and click View Source.

Visual Basic Editor opens and you see the button's empty click event procedure. You'll write code to get the map document, its page layout, and the selected element in the graphics container.

10 In the SetElementName click event procedure, declare and set a variable to get the map document.

```
Dim pMxDoc As IMxDocument
Set pMxDoc = ThisDocument
```

11 Declare and set a variable to get the page layout.

```
Dim pLayout As IPageLayout
Set pLayout = pMxDoc.PageLayout
```

As you name elements, you want to be sure that only one is selected at a time. You'll use the ElementSelectionCount property on IGraphicsContainerSelect to find out how many elements are selected. Then you'll write an If Then statement so your code runs only when the count is equal to 1.

12 Declare and set a variable to get the layout's IGraphicsContainerSelect interface.

Here you are switching interfaces from IPageLayout to IGraphicsContainerSelect.

```
Dim pGraphics As IGraphicsContainerSelect
Set pGraphics = pLayout
```

13 Start an If Then statement to determine how many elements are selected.

```
If pGraphics.ElementSelectionCount <> 1 Then

End If
```

14 Inside the If Then statement, use a message box to warn that only one element should be selected. Then use Exit Sub to exit the procedure.

```
MsgBox "Select one element"
Exit Sub
```

Now you can write the code to set a name for the selected element.

The DominantElement property on IGraphicsContainerSelect returns the IElement interface of a selected element. This isn't the interface you want, however. You want IElementProperties2 for its Name property.

Since all elements have IElementProperties2, and since you don't need IElement itself, you can declare your variable directly to IElementProperties2 and let VBA switch interfaces for you.

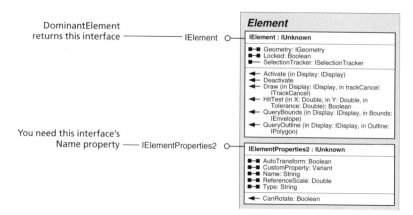

DominantElement returns this interface ——— IElement

You need this interface's Name property ——— IElementProperties2

15 After the If Then statement, declare and set a variable to get the selected element's IElementProperties2 interface using the DominantElement property.

```
Dim pElementProp As IElementProperties2
Set pElementProp = pGraphics.DominantElement
```

Next you will make an input box so you can type the selected element's name.

16 Declare a string variable and use an input box to set its value.

```
Dim strName As String
strName = InputBox("Enter a name", "Name the Graphic")
```

17 Use the string variable to set the selected element's name.

```
pElementProp.Name = strName
```

You are finished coding the SetElementName button. Now you'll code the GetElementName button, which is similar. In fact, to save time, you'll copy and paste most of the code.

Normally, you don't copy code from one procedure to another. As a rule, it's best to make one subroutine with common code that you then call from both click events. In the interest of finishing the exercise in the fewest number of steps, you will violate the rule.

18 In the SetElementName click event procedure, highlight the first twelve lines of code as shown below. Click the Edit menu and click Copy.

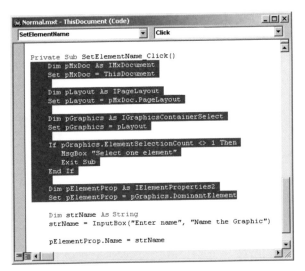

19 Click the object list drop-down arrow and click GetElementName.

The wrapper lines for the GetElementName button's click event procedure are added.

20 Click your mouse between the wrapper lines. Click the Edit menu and click Paste.

You will add a line of code to get the element's name and report it in a message box.

21 At the bottom of the click event procedure, after the pasted code, add the following line to report the element's name in a message box.

```
MsgBox "The element's name is: " & pElementProp.Name
```

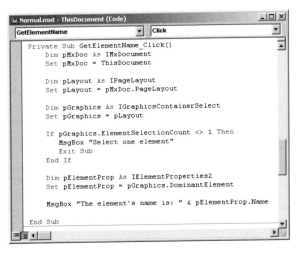

The two buttons are ready to test.

22 Close Visual Basic Editor.

23 On the Tools toolbar in ArcMap, click the Select Elements tool. On the layout page, click the map title to select it.

Map title —— Select the title

999 Toxic Sites

24 On the Layout toolbar, click SetElementName. Type **ToxicMapTitle**.

25 Click OK.

The text element representing the map title has now been named, which means it can be distinguished from other elements in the graphics container. You didn't change the title's text, which still reads "Map title." In the next exercise, you will associate the title's text with the user's choice in the Pick A State combo box.

Next, you will use the GetElementName button to confirm that the title's name has been set.

26 With the title selected, click GetElementName.

27 Click OK.

Now you will set the subtitle name.

28 On the layout page, click the map's subtitle (999 Toxic Sites) to select it. Click SetElementName and type **ToxicMapSubtitle**.

Map title

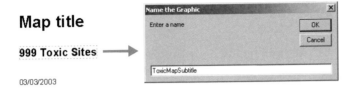

999 Toxic Sites

03/03/2003

29 Click OK.

Finally, you will set the text element representing the date.

30 On the layout page, select the map's date. Click SetElementName and type **ToxicMapDate**.

31 Click OK.

You have selected three elements and set their names. In the next exercise, you will write code to get these elements by name and update their text.

Since the buttons you created and the code you wrote for them are stored in your normal template, there is no need to save this map document in order to save the buttons and code. When working with the normal template, your edits are saved directly to the Normal.mxt file. There is no Save Normal button to click.

After creating the buttons, however, you used them to set the Name property of the three text elements. The text elements are part of the map document. To save those names you have to save the map document.

32 If you want to save your work, click the File menu in ArcMap and click Save As. Navigate to **C:\ArcObjects\Chapter19**. Rename the file **my_ex19a.mxd** and click Save. If you are continuing with the next exercise, leave ArcMap open. Otherwise close it.

If you no longer want the two UIButtons to appear, you can remove them by opening the Customize dialog box and dragging them off the Layout toolbar. To remove them completely from your Normal.mxt file, use the Delete UIControl button on the Commands tab of the Customize dialog box. You should complete exercise 19b before doing this, however, just in case you need to get or set the element names again.

Manipulating text elements

The application you have been building over the last three chapters is almost done. In the last exercise, you created two buttons to get and set the names of the three text elements you want to update. In this exercise, you will get those elements by looping through the graphics container and checking each element's Name property. After getting the map's title and subtitle, you'll change them to match the user's combo box selection. You'll also change the date to reflect the current date.

You loop through the elements on a layout using the IGraphicsContainer interface. Its Next and Reset methods work just as they do with an Enum.

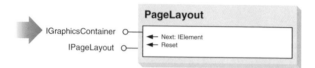

You get the IGraphicsContainer interface by using the IMxDocument interface's PageLayout property as you did in step 11 of exercise 19a. The PageLayout property, however, returns IPageLayout and you need IGraphicsContainer. That's OK, because VBA can do the QI for you, as shown with the code below.

```
Dim pMxDoc As IMxDocument
Set pMxDoc = ThisDocument

Dim pGraphics As IGraphicsContainer
Set pGraphics = pMxDoc.PageLayout
```

You can now get elements from the graphics container using the Next method. The first Next returns the first element's IElement interface. You don't need IElement, but you do need the Name property on IElementProperties2. Just as in the last exercise, you can declare a variable to IElementProperties2 and let VBA do QueryInterface for you.

```
Dim pElementProp As IElementProperties2
Set pElementProp = pGraphics.Next
```

When you get an element, you check its name. The following line of code uses an If Then statement to see if the element is the map title text:

```
If pElementProp.Name = "ToxicMapTitle" Then
```

Since you have three different elements to get, you'll use a Case statement in this exercise rather than an If Then statement.

Each of the elements you want to update comes from the TextElement class, shown below. The ITextElement interface has a Text property for setting text.

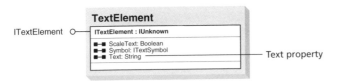

```
              TextElement
ITextElement O──  ITextElement : IUnknown

                 ■-■ ScaleText: Boolean
                 ■-■ Symbol: ITextSymbol
                 ■-■ Text: String ──────────── Text property
```

When you find an element you want, such as the map title, you switch interfaces from IElementProperties2 to ITextElement.

```
Dim pTextElement As ITextElement
Set pTextElement = pElementProp
```

Then you can set the Text property to a string or variable.

```
pTextElem.Text = "Rhode Island"
```

Rhode Island

31 Toxic Sites

7/16/2003 4:39:26 PM

Exercise 19b

In this exercise, you will use the element names you set in the last exercise to get the map's title, subtitle, and date. Then you'll write code that sets the title to the name of the U.S. state selected by the user, the subtitle to the number of toxic sites in the state, and the date to the current date. Your code, like the code you wrote in the previous two chapters, goes in the combo box's SelectionChange event procedure.

1 Start ArcMap and open **ex19b.mxd** in the **C:\ArcObjects\Chapter19** folder.

When ArcMap opens, you see the toxic sites layout.

2 Open the Customize dialog box.

3 On the Make a map toolbar, right-click the Pick a State combo box and click View Source.

You see the ThisDocument code module and the combo box's SelectionChange event procedure.

4 Scroll toward the bottom of the procedure. At the end of the If Then statement, locate the line of code that sets the extent of the active view for the EPA map.

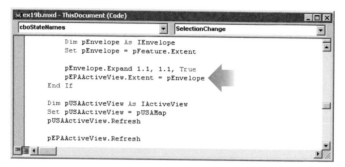

After this line, you will write code to get the layout's IGraphicsContainer interface. The PageLayout property returns IPageLayout, so you'll let VBA QI for you to get IGraphicsContainer.

5 Declare and set a variable to get the layout's IGraphicsContainer interface.

```
Dim pGraphics As IGraphicsContainer
Set pGraphics = pMxDoc.PageLayout
```

You'll write a Do Until loop and use the Next method on IGraphicsContainer to move through the elements on the layout page. You'll check the Name property of each element, and when you find the names you set in the last exercise, you will get those elements and update their Text property.

Before coding the loop, however, you have a couple of things to do. First, you'll reset the graphics container's pointer. Elements on a layout page are always being selected and changing their front-to-back position, and this affects which element is being pointed at. Although you haven't rearranged any elements, and the pointer is probably at the top of the list already, it's a good idea to make sure.

6 Reset the graphics container's pointer to the top of the list.

```
pGraphics.Reset
```

Now that the pointer is reset, you'll run the Next method one time to get the first element in the container. The Next method returns the IElement interface, but you want IElementProperties2, which has the Name property. As in the last exercise, you'll declare a variable to IElementProperties2, and let VBA do QueryInterface for you.

7 Declare and set a variable for the first element in the graphics container.

```
Dim pElementProp As IElementProperties2
Set pElementProp = pGraphics.Next
```

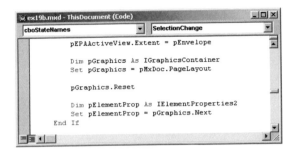

```
ex19b.mxd - ThisDocument (Code)
cboStateNames                    SelectionChange

        pEPAActiveView.Extent = pEnvelope

    Dim pGraphics As IGraphicsContainer
    Set pGraphics = pMxDoc.PageLayout

    pGraphics.Reset

    Dim pElementProp As IElementProperties2
    Set pElementProp = pGraphics.Next
End If
```

You now have the first element in the graphics container. If that element is one of the three you are looking for, you will set its Text property on the ITextElement interface. You will declare a variable to ITextElement here and switch to this interface inside the Do Until loop when you need it.

8 Declare a variable to the ITextElement interface.

```
Dim pTextElement As ITextElement
```

Now that you have variables pointing to the two interfaces you need, you can code the looping statement. The loop will test the pElementProp variable to see if it does indeed contain an element. If it contains Nothing, that means that the pointer has moved past the last element in the graphics container.

9 Start a Do Until loop that tests for Nothing.

```
Do Until pElementProp Is Nothing

Loop
```

10 Inside the loop, start a Case statement to see if the element is either the map's title, subtitle, or date.

```
Select Case pElementProp.Name

End Select
```

You may recall from chapter 5 that Case statements are used in multiple-choice situations. There, you used a Case statement to apply different tax rates to property types. Here, you will assign different text strings to elements depending on their names. Since you have three elements with names (and you don't care about the rest), your statement will have three cases.

11 Inside the Case statement, add a case to test for the map title. When the element is the map title, switch to the element's ITextElement interface and set its Text property equal to the combo box's EditText property.

```
Case "ToxicMapTitle"
    Set pTextElement = pElementProp
    pTextElement.Text = cboStateNames.EditText
```

The combo box's EditText property holds the name of whichever U.S. state the user picks. The code above assigns this name to the map title's Text property.

12 Add a second case to test for the map's subtitle. Set its Text property to the Selection Set count.

```
Case "ToxicMapSubtitle"
    Set pTextElement = pElementProp
    pTextElement.Text = _
        pSelectionSet.Count & " Toxic Sites"
```

In the previous chapter, you created a selection set of toxic sites and used its Count property to display the number of sites in a message box. The code above uses that same Count property to set the subtitle's text.

13 Add a third case to test for the map's date. Set its Text property to today's date with VBA's Now function, which returns the current date.

```
Case "ToxicMapDate"
    Set pTextElement = pElementProp
    pTextElement.Text = Now
```

Back in chapter 2, you used the Now function in the title bar of a message box.

14 Outside the Case statement (after the End Select line, but before the Loop keyword), add a Next method to move to the next element in the graphics container.

```
Set pElementProp = pGraphics.Next
```

You now need to refresh the layout. You could refresh the entire layout page, but it will be more efficient to use IActiveView's PartialRefresh method, which you used before in chapter 12.

You need the page layout's IActiveView interface. Since you already have a variable pointing to the page layout's IGraphicsContainer interface (from step 5), you can switch interfaces.

15 Immediately after the Loop keyword but before the End If statement, add the following code to get the page layout's active view and use its PartialRefresh method.

```
Dim pActiveView As IActiveView
Set pActiveView = pGraphics

pActiveView.PartialRefresh _
    esriViewGraphics, Nothing, Nothing
```

The arguments specify that all graphic elements on the layout page (but no other elements) will be refreshed. There aren't many graphic elements on this layout, so the redraw speed isn't an issue. However, when it is, you can refresh just one graphic. To do that, you specify the element as PartialRefresh's second argument: pActiveView.PartialRefresh esriViewGraphics, pElement, Nothing. You could get the ActiveView before the Case statement, and then inside each case, do a PartialRefresh on each element.

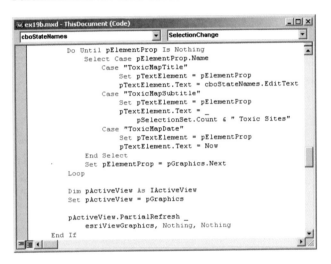

```
ex19b.mxd - ThisDocument (Code)

cboStateNames          SelectionChange

        Do Until pElementProp Is Nothing
            Select Case pElementProp.Name
                Case "ToxicMapTitle"
                    Set pTextElement = pElementProp
                    pTextElement.Text = cboStateNames.EditText
                Case "ToxicMapSubtitle"
                    Set pTextElement = pElementProp
                    pTextElement.Text = _
                        pSelectionSet.Count & " Toxic Sites"
                Case "ToxicMapDate"
                    Set pTextElement = pElementProp
                    pTextElement.Text = Now
            End Select
            Set pElementProp = pGraphics.Next
        Loop

        Dim pActiveView As IActiveView
        Set pActiveView = pGraphics

        pActiveView.PartialRefresh _
            esriViewGraphics, Nothing, Nothing
    End If
```

The code is almost ready to test. In the previous chapter, you used a message box to report the number of toxic sites in a state. Now your code puts that number right on the map, so you'll comment out the message box line.

16 Scroll up through the code until you find the message box that reports the number of toxic sites.

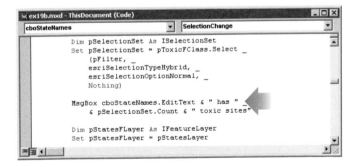

```
ex19b.mxd - ThisDocument (Code)

cboStateNames          SelectionChange

        Dim pSelectionSet As ISelectionSet
        Set pSelectionSet = pToxicFClass.Select _
            (pFilter, _
            esriSelectionTypeHybrid, _
            esriSelectionOptionNormal, _
            Nothing)

        MsgBox cboStateNames.EditText & " has " _
            & pSelectionSet.Count & " toxic sites"

        Dim pStatesFLayer As IFeatureLayer
        Set pStatesFLayer = pStatesLayer
```

chapter

13
14
15
16
17
18
19
20

17 Put a single quote in front of the line to comment it out.

```
'MsgBox cboStateNames.EditText & " has " _
    & pSelectionSet.Count & " toxic sites"
```

18 Close Visual Basic Editor.

19 On the Make a map toolbar, click the Pick a State drop-down arrow and click Arizona.

The detail and overview maps update. In the upper left corner of the map, the title, subtitle, and date change as well. You'll zoom in to see this better.

20 On the Layout toolbar (not the Tools toolbar), click the Zoom In tool. Zoom in on the upper left corner of the map.

Arizona

140 Toxic Sites

4/15/2003 10:52:13 AM

21 If you want to save your work, click the File menu in ArcMap and click Save As. Navigate to **C:\ArcObjects\Chapter19**. Rename the file **my_ex19b.mxd** and click Save. If you are continuing with the next chapter, leave ArcMap open. Otherwise close it.

Editing tables

Adding fields
Getting and setting values

The features you manage in a GIS, such as land parcels, cities, streets, crime scenes, and customer addresses, appear as rows (records) in a table. The columns (fields) in the table represent categories of information. A land parcel table, for example, has a unique record for every parcel in a data set. It has fields for things like the parcel size, the address, the owner's name, and the zoning code. The intersection of a record and a field is a cell. A cell holds a particular piece of information (a value) about a record.

In this chapter, you will edit cell values. To edit a cell value, you get an existing value from a table and set it to the new value you want. This process involves making a cursor, moving its pointer to a particular record, and specifying a particular field. You know from chapter 18 how to make cursors and move their pointers. Fields are specified by their index position numbers. The first field in a table has position 0, the second field has position 1, and so on.

The graphic below illustrates the idea of getting a cell value from a cursor. A variable, pFeature, is set up to point at a record, and a field's index position is specified. The combination of the first record and the fifth field identifies a unique cell value—in this case, "Florin."

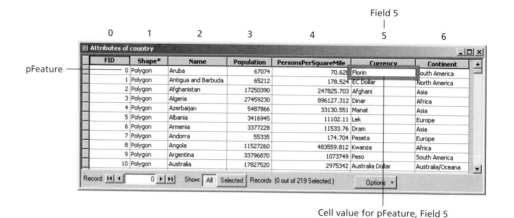

Cell value for pFeature, Field 5

In the first exercise of this chapter, you will add a new field to a table. In the second exercise, you will get cell values from two existing fields and use them to calculate values for the new field.

Adding fields

The classes on the diagram below, with the exception of Field and Fields, should be familiar to you. The diagram shows that a feature class has a fields object, which is a collection of its fields. You have worked with collection objects before. For example, in chapter 17, you used the maps collection object to get the EPA and USA maps by their index numbers. The fields object serves the same purpose for fields: it gives you a way to access fields in a feature class, in this case, with a method called FindField.

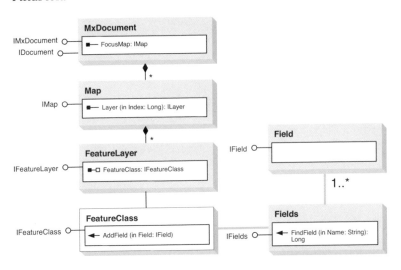

To add a field to a feature class (which is a specific type of table), you start by creating a new field from the Field coclass with the standard two lines of code.

```
Dim pField As IField
Set pField = New Field
```

Next, you set the field's properties. You have to do that before adding the field to the table, because once the field has been added, its properties can't be changed.

The following diagram shows that the Field coclass has two nearly identical interfaces: IField and IFieldEdit. The difference between them is that IField has only left-hand barbells, so you can get properties but not set them, and IFieldEdit has only right-hand barbells, so you can set properties but not get them. The interfaces are designed to keep the actions of getting and setting properties separate. (But if you remember interface inheritance, you will see a loophole: IFieldEdit inherits from IField.)

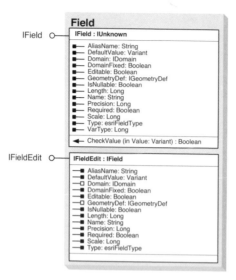

Field

IField ○—— | **IField : IUnknown** |
| :--- |
| ■— AliasName: String |
| ■— DefaultValue: Variant |
| ■— Domain: IDomain |
| ■— DomainFixed: Boolean |
| ■— Editable: Boolean |
| ■— GeometryDef: IGeometryDef |
| ■— IsNullable: Boolean |
| ■— Length: Long |
| ■— Name: String |
| ■— Precision: Long |
| ■— Required: Boolean |
| ■— Scale: Long |
| ■— Type: esriFieldType |
| ■— VarType: Long |
| ◄— CheckValue (in Value: Variant) : Boolean |

IFieldEdit ○—— | **IFieldEdit : IField** |
| :--- |
| —■ AliasName: String |
| —■ DefaultValue: Variant |
| —□ Domain: IDomain |
| —■ DomainFixed: Boolean |
| —■ Editable: Boolean |
| —□ GeometryDef: IGeometryDef |
| —■ IsNullable: Boolean |
| —■ Length: Long |
| —■ Name: String |
| —■ Precision: Long |
| —■ Required: Boolean |
| —■ Scale: Long |
| —■ Type: esriFieldType |

After creating a field, you switch to its IFieldEdit interface to set properties. (You could also have declared a variable directly to IFieldEdit when you created the new field.)

```
Dim pFieldEdit As IFieldEdit
Set pFieldEdit = pField
```

Fields have many properties, but you don't always need to set all or even most of them. The two that are essential are the name and the data type. You set the field's Name property equal to a string.

```
pFieldEdit.Name = "Population"
```

The data type determines what kind of values the field will hold—strings, numbers, dates, and so on. You set the Type property with esriFieldType constants like esriFieldTypeDate, esriFieldTypeString, or esriFieldTypeInteger. You can find a complete list of these constants in the developer help.

```
pFieldEdit.Type = esriFieldTypeInteger
```

At this point, you could go ahead and add the field to the table. It's a good idea, though, to make sure that a field with the same name doesn't already exist. Here's where the Fields collection object and its FindField method are useful. To get the fields collection, you use the Fields property on the IFeatureClass interface. For the moment, you'll skip over the code that gets the feature class (which you learned how to write in chapter 18) and just assume that you have a pFClass variable pointing to IFeatureClass. You then get the fields collection with the following code:

```
Dim pFields As IFields
Set pFields = pFClass.Fields
```

The FindField method on IFields takes a field name as its argument and searches for this field in the collection. If it finds it, it returns the field's index position number. If the field isn't there, it returns the value –1. The code below declares an integer variable to hold the returned value and searches the fields collection for a field called Population.

```
Dim intPosPopField As Integer
intPosPopField = pFields.FindField("Population")
```

You can write an If Then statement that tests for the returned value. If the value is –1, you can add the new field to the map with full confidence that it's not already there.

```
If intPosPopField = -1 Then

End If
```

To add the field, you run the AddField method on IFeatureClass. Your pFClass variable (the one you are assuming you already have) points to this interface.

```
pFClass.AddField pField
```

In the exercise, you will run code like this from a UIButton, which you might call the Add Field button. (Actually, you'll give it a different name in the exercise, because it will end up doing a lot more than just adding a field, but that's all it will do at first.)

If you wanted to add your new field to a specific feature class, you might put the button on a toolbar or menu, and code its click event to get the feature class you want from the layer you want.

But say that you want the code to work on any feature layer in the map. That calls for a different strategy. One thing you could do is code the Add Field button's enabled event procedure with some If Then and TypeOf statements to gray out the button unless a feature layer was selected in the table of contents.

A better solution, however, is to put the button on the feature layer context menu. Unlike a toolbar or menu, a context menu is not always available from the interface—it only appears when the user has right-clicked an object. Putting the Add Field button on the context menu means you don't have to worry about how and when to enable it. The button will be available only when a feature layer has been right-clicked.

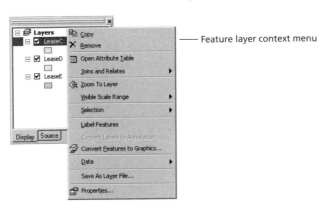

———— Feature layer context menu

Adding UIControls to toolbars and menus is easy for you by now. Adding a UIButton to a context menu is a little different because of the way context menus work—you have to be right-clicking the mouse to see them. But since you have to be left-clicking to drag a control from the Customize dialog box, you can't drag a control to a context menu.

ESRI programmers have designed a way around this, of course. A copy of every context menu is stored on a special toolbar called the Context Menu toolbar. When you want to add a control to a context menu, you add it to this toolbar. ArcMap then automatically takes care of putting it on the correct context menu for you.

Context menus might seem a little magical in that whenever you right-click an object on the interface, the appropriate context menu appears. But it's not magic, just logic. ArcMap has been programmed to know where you right-click on the screen. It gets the screen coordinates of your click, uses them to deduce which object you clicked on, and displays that object's context menu.

Since ArcMap knows which object you've right-clicked on, it seems reasonable that you could ask for that object. You can. The ContextItem property on IMxDocument returns whichever object a user has right-clicked on. This gives you a convenient way to work with objects that the user selects and spares you the trouble of trying to anticipate and control their choices in your code.

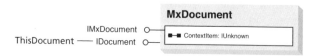

The ContextItem property returns an object's IUnknown interface. Why IUnknown? Because on the one hand, ContextItem can get a wide variety of objects—layers, maps, active views, points, lines, and polygons—depending on where the user right-clicks. But on the other hand, programming logic requires that all properties and methods return a single, specified interface. IUnknown is used in this situation because it's the one interface that all objects share.

So far (in theory), you have made an Add Field button and put it on the feature layer context menu. When the user right-clicks a feature layer and clicks this button, code in the button's click event procedure runs. You have seen the part of the code that makes the new field, sets its properties, checks to make sure that the field doesn't already exist, and adds the field to the layer's feature class.

You also assumed that you had already gotten the feature class and had a variable pointing to IFeatureClass. Now you'll drop this assumption and look at the code that actually gets the feature class using the ContextItem property you just learned about.

First, you get the IMxDocument interface, because it has the ContextItem property.

```
Dim pMxDoc As IMxDocument
Set pMxDoc = ThisDocument
```

Then you declare a variable to IUnknown and set it with ContextItem.

```
Dim pUnknown As IUnknown
Set pUnknown = pMxDoc.ContextItem
```

You know that pUnknown points to a feature layer. How? Because the Add Field button's code—the code you're looking at now—can't be running unless the button has been clicked. The button can't be clicked, however, unless the context menu that it's located on has been opened. But the context menu only opens when a feature layer has been right-clicked. So if the code is executing, and the ContextItem property has returned an object, the object has to be a feature layer.

Since you are pointing at the IUnknown interface of a feature layer, you can do QueryInterface to get its IFeatureLayer interface.

```
Dim pFLayer As IFeatureLayer
Set pFLayer = pUnknown
```

Alternatively, you could shorten the last four lines to two by letting VBA do the QueryInterface for you.

```
Dim pFLayer As IFeatureLayer
Set pFLayer = pMxDoc.ContextItem
```

Finally, you get the feature class from the layer with the FeatureClass property.

```
Dim pFClass As IFeatureClass
Set pFClass = pFLayer.FeatureClass
```

Exercise 20a

Foresters group areas of similar trees into units called stands, which are assessed and given a dollar value. A healthy stand, for example, might be worth sixty dollars or more per square meter. Stands, in turn, are aggregated into larger entities called leases. The right to harvest a lease is auctioned by landowners to the highest bidder. When a lease goes out to bid, GIS data describing the stands is distributed to the bidders.

A stand within a lease

You are a programmer for a lumber company that bids on leases for the right to harvest trees. When you get a map layer of the lease, it contains a record for each stand. The attribute table has two fields: ValuePerMeter and Shape_Area (the stand's area in square meters). Multiplying the values in these two fields gives you the stand's total value.

In the graphic below, the first stand has a value of $64 per square meter. Since its area is 15,680 square meters, the stand is worth just over a million dollars.

Value per meter   Area

ValuePerMeter	StandID	Shape_Length	Shape_Area
64.000000	6224	769.689245	15680.172005
42.000000	1164	2164.996611	175721.964087
63.000000	6223	650.045632	14867.984924
30.000000	1169	5261.256657	352276.409512
63.000000	1171	776.790314	31344.562658
63.000000	1178	1307.944741	67274.578893
53.000000	1198	1768.194060	95401.656306
52.000000	1209	946.578676	52427.342690
53.000000	1216	714.853307	27774.296188
25.000000	1229	4421.968632	349366.263451

Record: 0   Show: All Selected   Records (0 out of 1405 Selected.)   Options

The bidders at your company need to know not just the values of individual stands, but the total lease value, which is the sum of the stand values. Since this is a routine you go through every time a new lease hits your desk (or hard drive), you have decided to automate the task.

In this exercise, you will create a UIButton and add it to the feature layer context menu. That way, your code will work on any feature layer in the map document, which will be useful if there are multiple leases to evaluate. Then you will write code in the UIButton's click event procedure to add a new field to the layer's attribute table. The field will eventually hold the dollar value of each stand.

In the second exercise, you'll write the code that calculates the value of each stand, totals these values into a lease value, and reports the lease value in a message box.

1 Start ArcMap and open **ex20a.mxd** in the **C:\ArcObjects\Chapter20** folder.

The map opens and you see three layers: LeaseC, LeaseD, and LeaseE.

2 Open the Customize dialog box. Click the Commands tab. In the Categories list, click UIControls.

3 Make sure the Save in drop-down list is set to ex20a.mxd. Click New UIControl and click Create.

In the list of commands, you see the new control. (If you haven't removed them, you also see the GetElementName and SetElementName controls you made in the last chapter.)

4 Rename the new UIButton **Project.LeaseValue**. Press Enter.

Now you'll drag the new UIButton to the Context Menus toolbar.

5 In the Customize dialog box, click the Toolbars tab. Check the box next to the Context Menus toolbar.

The Context Menus toolbar displays.

6 If necessary, move the Customize dialog box so it doesn't overlap the Context Menus toolbar.

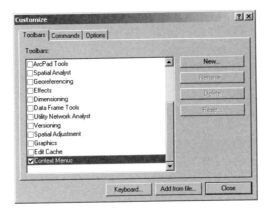

In the next step, you'll drag the button to the Context Menu toolbar. As you do so, you will see a list of every ArcMap context menu. The Feature Layer Context Menu is about halfway down, and you may have to scroll down to see it.

7 Click the Commands tab and drag the LeaseValue button to the Context Menu toolbar. Drag down until you see the Feature Layer Context Menu, then drop the button under the Open Attribute Table choice.

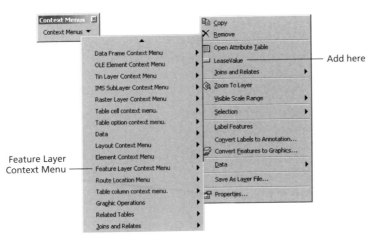

Feature Layer
Context Menu

Add here

Adding fields

8 On the Feature Layer Context Menu, right-click LeaseValue and click Text Only. Right-click again and change the text to **Report Lease Value**. Press Enter.

Button names can't have spaces, and as far as VBA is concerned, the button's name is still LeaseValue. Your users, however, will see the more descriptive words Report Lease Value on the context menu.

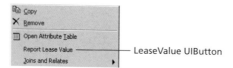

LeaseValue UIButton

Next you will write the code for this button.

9 Right-click Report Lease Value and click View Source.

As Visual Basic Editor opens, the Context Menu toolbar automatically closes. (This toolbar is available only when the Customize dialog box is open.) You see the empty LeaseValue click event procedure.

In the next three steps, you will set up variables to get the map document, the feature layer, and the layer's feature class.

10 In the click event procedure, declare and set a variable for the map document.

```
Dim pMxDoc As IMxDocument
Set pMxDoc = ThisDocument
```

You want to get the feature layer that the user right-clicks so that you can get its feature class. The ContextItem property on IMxDocument gets this layer's IUnknown interface, but you need IFeatureLayer, which has a property to get the feature class.

You'll declare a variable to IFeatureLayer, and let VBA switch interfaces from IUnknown to IFeatureLayer for you.

11 Declare an IFeatureLayer variable and set it with the ContextItem property.

```
Dim pFLayer As IFeatureLayer
Set pFLayer = pMxDoc.ContextItem
```

The FeatureClass property on IFeatureLayer returns the layer's feature class.

12 Declare and set a variable for the layer's feature class.

```
Dim pFClass As IFeatureClass
Set pFClass = pFLayer.FeatureClass
```

Before adding a field to the feature class, you'll make sure that a field with the same name doesn't already exist. To do that, you'll get the feature class's Fields collection, which has the FindField method on its IFields interface.

13 Declare and set a variable for the Fields collection.

```
Dim pFields As IFields
Set pFields = pFClass.Fields
```

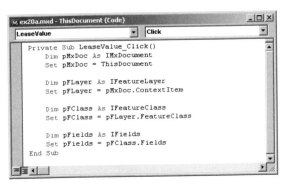

FindField takes a field name as its argument and returns the field's index position. If the field isn't found, a value of −1 is returned.

The field name you'll search for is StandValue, because this is going to be the name of the new field you add. Obviously, this field will not be found the first time the code runs (since you haven't created it yet), but you are guarding against future possibilities.

14 Declare and set an integer variable to hold the StandValue field's index position number.

```
Dim intPosStandValue As Integer
intPosStandValue = pFields.FindField("StandValue")
```

15 Begin an If Then statement to check whether the field exists.

```
If intPosStandValue = -1 Then

End If
```

After confirming that there is no field with that name, you'll write code inside the If Then statement to create the field. When creating a field, you can declare its variable to either IField or IFieldEdit. Since you can't set a field's properties with the IField interface, you will use IFieldEdit.

16 Inside the If Then statement, declare and set a variable to create a new field.

```
Dim pFieldEdit As IFieldEdit
Set pFieldEdit = New Field
```

Now you will set the field's Name and Type properties and add it to the feature class.

17 Set the field's Name and Type properties.

```
pFieldEdit.Name = "StandValue"
pFieldEdit.Type = esriFieldTypeDouble
```

18 Use the AddField method to add the field to the table.

```
pFClass.AddField pFieldEdit
```

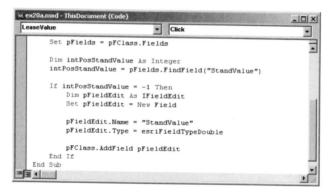

The code is ready to test.

19 Close Visual Basic Editor.

20 In the table of contents, right-click the LeaseC layer, and click Open Attribute Table. Scroll all the way to the right side of the table.

The StandValue field will be added here

ValuePerMeter	StandID	Shape_Length	Shape_Area
64.000000	6224	769.689245	15680.172005
42.000000	1164	2164.996611	175721.964087
63.000000	6223	650.045632	14867.984924
30.000000	1169	5261.256657	352276.409512
63.000000	1171	776.790314	31344.562658
63.000000	1178	1307.944741	67274.578893
53.000000	1198	1768.194060	95401.656306
52.000000	1209	946.578676	52427.342690
53.000000	1216	714.853307	27774.296188
25.000000	1229	4421.968632	349366.263451

Record: 0   Show: All Selected   Records (0 out of 1405 Selected.)   Options

21 If necessary, move the table out of the way. In the table of contents, right-click the LeaseC layer and click Report Lease Value.

The StandValue field is added to the table. By default, all the cell values are Null. In the next exercise, you will set these values by multiplying each record's ValuePerMeter by its Shape_Area.

New field added

OBJECTID	Shape*	ValuePerMeter	StandID	Shape_Length	Shape_Area	StandValue
1	Polygon	32	93	3062.866195	158705.375680	<Null>
2	Polygon	52	115	1619.680690	114920.218270	<Null>
3	Polygon	32	121	1745.766785	160169.268034	<Null>
4	Polygon	53	126	1250.656780	47705.984241	<Null>
5	Polygon	12	124	6336.713968	418949.487108	<Null>
6	Polygon	26	128	1138.476359	44586.813107	<Null>
7	Polygon	31	133	2572.347274	217903.829622	<Null>
8	Polygon	31	4892	638.283178	11271.638917	<Null>
9	Polygon	52	146	2180.264852	123314.403526	<Null>

Attributes of LeaseC

Record: 0   Show: All Selected   Records (0 out of 241 Selected.)   Options ▾

22 Close the table.

23 If you want to save your work, click the File menu in ArcMap and click Save As. Navigate to **C:\ArcObjects\Chapter20**. Rename the file **my_ex20a.mxd** and click Save. If you are continuing with the next exercise, leave ArcMap open. Otherwise close it.

Saving an exercise map document with a new name usually means that you can reopen the original document and do the exercise again whenever you like. In this case, however, you have changed a feature class (by adding a field to it), and this change is independent of the map document. If you open ex20a.mxd again, the LeaseC attribute table will already contain the StandValue field. You can redo the exercise by manually deleting this field from the table.

Getting and setting values

In this exercise, you have two main tasks. First, you will replace all the Null values in the StandValue field with calculated values. This means looping through a feature cursor to set the cell value for each stand in turn. Second, you will sum the stand values to come up with a lease value. This means looping through each feature again with a second cursor to get all the stand values and add them up.

In chapter 18, you used the Search method to make a cursor for reading values in a table. In this exercise, however, you need to change values, so you will use the Update method instead. Both methods require a feature class and a query filter to create a cursor. You already have code from the last exercise that gets the feature class. As for the query filter, you are not actually going to make one (even though you know how). That's because when you want a cursor to get every feature, you use the Nothing keyword instead of a filter. The code below declares an IFeatureCursor variable and sets it with the Update method.

```
Dim pFCursor As IFeatureCursor
Set pFCursor = pFClass.Update(Nothing, False)
```

After creating a feature cursor, you move through it with the NextFeature method.

```
Dim pFeature As IFeature
Set pFeature = pFCursor.NextFeature
```

The part that's new for you is the process of getting and setting cell values. The first thing you do is move the pointer to the desired feature. Say that you want to get the population value for Andorra in the example below. Since Andorra is the eighth record in the table, you use the NextFeature method eight times (probably in a loop) to get there.

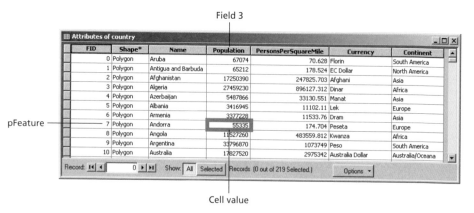

Field 3

pFeature

Cell value

To get a particular cell value for this record, you use a Value property. As shown in the simplified geodatabase diagram below, the Value property is found on the IRowBuffer interface of the RowBuffer class.

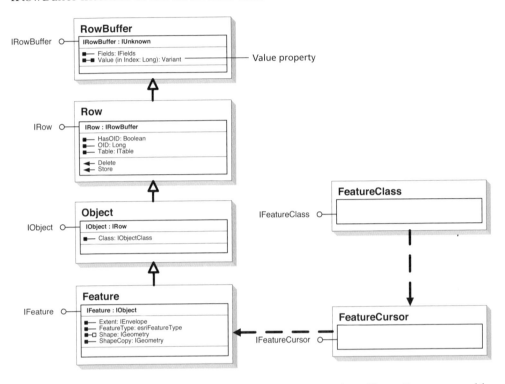

At first glance, this might seem like a place for QueryInterface. Your pFeature variable points to IFeature and the Value property you need is on an interface several classes up in the hierarchy. But as it turns out, there is a chain of interface inheritance that goes from IFeature all the way up to IRowBuffer, which means you can use the Value property just as if it were on IFeature.

The Value property takes a field's index position for its argument and returns a cell value. Say that you want to get Andorra's population. The Population field (in the table above) is at index position 3. The following line of code gets this value and displays it in a message box:

```
Msgbox "Andorra's population is: " & pFeature.Value(3)
```

The Value property has left and right barbells, so you can set it just as easily as you can get it. Say that Andorra's population has grown to 60,000. You write the following line of code to update that value:

```
pFeature.Value(3) = 60000
```

Although you are only working with features in this exercise, you may be curious about the Object, Row, and RowBuffer classes in the previous diagram. The triangles below each class mean that a feature is a type of object, an object is a type of row, and a row is a type of row buffer.

Objects are like features, except they are nonspatial (they have no Shape field). Both objects and features in a geodatabase can have subtypes and participate in geodatabase relationship classes. Rows represent records in a table at a generic level without the aforementioned geodatabase functionality.

RowBuffer has an interesting name. To GIS people, a buffer is a zone around a feature, but to programmers it means a location in memory. A row buffer is a record (with all its attributes) that is stored somewhere in your computer's active memory, but not in a table. Because of the inheritance hierarchy described above, features are row buffers and as such are stored in active memory.

This marks a milestone in your learning. The word "feature" has two meanings. It has the general meaning of a location represented by geometry (for example, a land parcel may be represented by a polygon feature or a city by a point feature). Every feature, in this sense, has a corresponding record in the layer's attribute table. For programmers, feature has a different meaning. It is an object returned by a feature cursor; in other words, a type of row buffer object that only exists in memory.

It's natural to talk about the pointer in a feature cursor as if it points at features in a table. In reality, however, it points at the programmer's kind of feature that is temporarily stored in memory. Each time you move the pointer with the NextFeature method, a feature is created in memory. Attribute values are copied into it from the corresponding record in the layer attribute table. Internally, a link is set up so the in-memory feature knows where it came from in the table.

You might think that all this creating of features in memory and copying of attribute values would slow your code down, but the effect is just the opposite. Actions on features in active memory happen much faster than actions on features stored in a table on a computer's hard disk.

When you set a feature's Value property, therefore, you are setting the value of the in-memory feature, not the actual row in the table. This means that there is a need for one more step. You have to tell the in-memory feature to write its values back to the corresponding feature in the table. To do that, you use the UpdateFeature method on the IFeatureCursor interface.

The code below updates pFeature's values in the database table.

```
pFCursor.UpdateFeature pFeature
```

You now have an idea how to get and set cell values. When you want to process a number of records, you write a Do Until loop, as you have done before with elements in a graphics container and symbols in an Enum. In the loop, you set values for the in-memory feature, update its corresponding feature in the feature class (table), and get the next feature.

```
Do Until pFeature Is Nothing
    'Do something to the feature
    pFeature.Value(3) = 60000

    'Write the feature's attributes to the table
    pFCursor.UpdateFeature pFeature

    'Get the next feature
    Set pFeature = pCursor.NextFeature
Loop
```

Exercise 20b

In the previous exercise, you wrote code to add the StandValue field to a feature class. In this exercise, you will write code to calculate the value of each stand. To do that, you'll multiply values in the ValuePerMeter field by values in the Shape_Area field and write the result to the StandValue field. You will use a feature cursor and a loop to repeat the process for every stand. Then you'll write code to total the stand values into a lease value (requiring another cursor and another loop). Finally, you'll convert the result into currency format, and report it in a message box.

OBJECTID	Shape*	ValuePerMeter	StandID	Shape_Length	Shape_Area	StandValue
1	Polygon	32	93	3062.866195	158705.375680	<Null>
2	Polygon	52	115	1619.680690	114920.218270	<Null>
3	Polygon	32	121	1745.766785	160169.268034	<Null>
4	Polygon	53	126	1250.656780	47705.984241	<Null>
5	Polygon	12	124	6336.713968	418949.487108	<Null>
6	Polygon	26	128	1138.476359	44586.813107	<Null>
7	Polygon	31	133	2572.347274	217903.829622	<Null>
8	Polygon	31	4892	638.283178	11271.638917	<Null>
9	Polygon	52	146	2180.264852	123314.403526	<Null>

Attributes of LeaseC

Record: 0   Show: All Selected Records (0 out of 241 Selected.)   Options

1 Start ArcMap and open **ex20b.mxd** in the **C:\ArcObjects\Chapter20** folder.

You see the map of leases. You will locate the LeaseValue click event procedure and add code to it.

2 Click the Tools menu, point to Macros, and click Visual Basic Editor.

Usually you get to source code by opening the Customize dialog box, right-clicking a control, and clicking View Source. You could do that here, too, but because of the way context menus work, it's cumbersome. (You have to display the Context Menu toolbar, scroll down to the Feature Layer context menu, and continue from there.) The alternative of opening Visual Basic Editor from the Tools menu is always available.

3 Make sure that the ThisDocument code module for the ex20b.mxd project is open. Bring it forward if necessary. Scroll to the bottom of the LeaseValue click event procedure and locate the End If line.

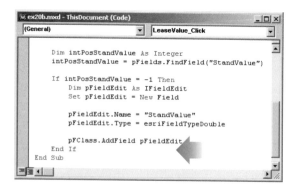

This is the end of the If Then statement that adds the StandValue field to the table. Before you can write the code to calculate each stand's value, you need the index positions of the StandValue, ValuePerMeter, and Shape_Area fields. You can get these numbers with the FindField method.

When the If Then statement runs, it means that the StandValue field doesn't exist and therefore the value stored in intPosStandValue is −1. Code in the If Then statement adds the new field. The new field is assigned a position, but the value in the intPosStandValue variable is still −1. So at the end of the If Then statement, you will write code to reset the intPosStandValue variable with the new field's position.

4 Immediately before the End If statement, add a line to reset the variable that contains the StandValue field's index position number. (If you want, you can copy the line from just before the If Then statement and paste it in the new location.)

```
intPosStandValue = pFields.FindField("StandValue")
```

Now you'll add code after the If Then statement to get the position numbers of the other two fields.

5 Immediately after the End If statement, declare and set an integer variable to hold the ValuePerMeter field's index position number.

```
Dim intPosValuePerMeter As Integer
intPosValuePerMeter = pFields.FindField("ValuePerMeter")
```

6 Declare and set an integer variable to hold the Shape_Area field's index position number.

```
Dim intPosShape_Area As Integer
intPosShape_Area = pFields.FindField("Shape_Area")
```

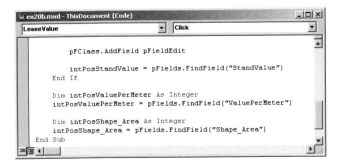

Next, you will create a feature cursor. In chapter 18, you used IFeatureClass's Search method to make a cursor. Here you will use the Update method (on the same interface) because you plan to edit the features.

You need a feature class and a query filter. You already got the feature class in the previous exercise, so you have a pFClass variable pointing to IFeatureClass. All you have to worry about now is the query filter—or do you?

Since every stand value is going to be calculated, you want the cursor to contain every feature in the feature class. When you want all features, you can use the keyword Nothing as the query filter. This means that no filter is applied and all features are included.

The Update method's second argument, like the Search method's, is for cursor recycling.

CURSOR RECYCLING

The recycling argument can be either true or false. When it is false (the cursor is nonrecycling), each time you run the NextFeature method, a new in-memory feature is created and stored in a location separate from the one before it.

Nonrecycling cursors let you hold many features in memory at the same time. You use them in situations where you are making simultaneous edits. For example, if you created a collection of adjacent land parcels that you wanted to offset a certain distance from the street, you would store each of those features in memory at the same time. Then you would apply a method (MoveSet on IFeatureEdit) to move them all at once.

When the argument is true (the cursor is recycling), a single feature is created in memory and reused. The first time you run NextFeature a new in-memory feature is created and the first row's data values are copied from the table to the feature (just like with a nonrecycling cursor). With each subsequent NextFeature, the next row's data values are copied to the same in-memory feature, overwriting the feature's previous values.

Recycling cursors can be used whenever you are processing a single feature at a time. They run faster than nonrecycling cursors because they require less memory.

In this case, since your code works on one feature at a time, you'll set the recycling argument to true. (False would also work, but it would take a little longer.)

7 Declare and set a variable to create a feature cursor.

```
Dim pFCursor As IFeatureCursor
Set pFCursor = pFClass.Update(Nothing, True)
```

8 Declare an IFeature variable and set it with the cursor's NextFeature method.

```
Dim pFeature As IFeature
Set pFeature = pFCursor.NextFeature
```

The cursor's pointer now points to the first feature.

Next, you'll start a Do Until loop to process each feature. The loop will end when the pointer moves past the last feature and points at Nothing.

9 Start a Do Until loop that runs until there are no more features in the cursor.

```
Do Until pFeature Is Nothing

Loop
```

In steps 4 through 6 you set up (and reset) variables to hold the index positions of the StandValue, ValuePerMeter, and Shape_Area fields. You will use these variables now as arguments for the Value property. Your code will get the cell value in each feature's ValuePerMeter field, multiply it by the cell value in Shape_Area, and use the result to set the cell value in StandValue.

10 Inside the loop, use the Value property to set the first feature's StandValue equal to the product of its ValuePerMeter and Shape_Area fields.

```
pFeature.Value(intPosStandValue) = _
    pFeature.Value(intPosValuePerMeter) * _
    pFeature.Value(intPosShape_Area)
```

So far, the new stand value is only held in memory. To write it to the feature class, you use the UpdateFeature method on IFeatureCursor. Your pFCursor variable from step 7 already points to this interface.

11 Use the UpdateFeature method to write the new value to the feature class.

```
pFCursor.UpdateFeature pFeature
```

The only thing missing from your Do Until loop is a line to move the pointer to the next feature in the cursor.

12 Use NextFeature to get the cursor's next feature.

```
Set pFeature = pFCursor.NextFeature
```

Without this line, you would have an endless loop. pFeature would always point to the first feature in the table and its stand value would be edited over and over again. (If you ever get stuck in an endless loop, press Ctrl + Break on the keyboard to get out of it.)

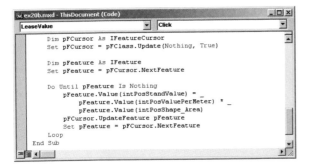

```
ex20b.mxd - ThisDocument (Code)
LeaseValue                          Click
    Dim pFCursor As IFeatureCursor
    Set pFCursor = pFClass.Update(Nothing, True)

    Dim pFeature As IFeature
    Set pFeature = pFCursor.NextFeature

    Do Until pFeature Is Nothing
        pFeature.Value(intPosStandValue) = _
            pFeature.Value(intPosValuePerMeter) * _
            pFeature.Value(intPosShape_Area)
        pFCursor.UpdateFeature pFeature
        Set pFeature = pFCursor.NextFeature
    Loop
End Sub
```

The first part of the code is ready to test.

EDITING FEATURE GEOMETRY

The Value property can be used to set any feature attribute value, including that of the Shape field. Say you want to move one of the features in a point feature class to a new location. Outside the feature class, you could create (or get) a point and set its x and y properties to the desired location. You could then use that point to set the value of the point feature you want to move. In the code below, the index position of the Shape field is assumed to be 1.

```
Dim pPoint As IPoint
Set pPoint - New Point
pPoint.X = 100
pPoint.Y = 200

pFeature.Value(1) = pPoint
```

If you are not sure of the Shape field's position, you can also set a feature's shape with the IFeature interface's Shape property.

```
pFeature.Shape = pPoint
```

13 Leave Visual Basic Editor open, but bring the ArcMap window forward.

Your If Then statement from the previous exercise checks whether a StandValue field already exists before it tries to add one. Thanks to this error checking, you can run your new code on a Lease layer whether or not the field has been added to it.

14 In the ArcMap table of contents, right-click the LeaseC layer and click Report Lease Value.

To see the new stand values, you need to open the attribute table.

15 In the ArcMap table of contents, right-click the LeaseC layer and click Open Attribute table. Scroll to the right if necessary.

New values here

OBJECTID	Shape*	ValuePerMeter	StandID	Shape_Length	Shape_Area	StandValue
1	Polygon	32	93	3062.866195	158705.375680	5078572.021748
2	Polygon	52	115	1619.680690	114920.218270	5975851.350039
3	Polygon	32	121	1745.766785	160169.268034	5125416.577100
4	Polygon	53	126	1250.656780	47705.984241	2528417.164764
5	Polygon	12	124	6336.713968	418949.487108	5027393.845301
6	Polygon	26	128	1138.476359	44586.813107	1159257.140784
7	Polygon	31	133	2572.347274	217903.829622	6755018.718281
8	Polygon	31	4892	638.283178	11271.638917	349420.806426
9	Polygon	52	146	2180.264852	123314.403526	6412348.983338

Record: 1 ▶ ▶I Show: All Selected Records (0 out of 241 Selected.) Options ▼

16 Close the table.

Now you will add code to sum the values in the StandValue field and report the total lease value.

17 Bring the Visual Basic Editor window forward. Scroll to the bottom of the LeaseValue click event procedure.

Your new code will go at the end of the procedure. You'll begin by declaring a variable to hold the sum of the stand values.

18 After the Loop keyword, declare a Double variable.

```
Dim dblTotal As Double
```

You have already created an Update cursor to set the stand values for each feature. Now you need a second cursor to get those values and total them. Since you aren't going to edit any cell values, you will use a Search cursor, as you did in chapter 18.

You don't have to declare any new variables—you can reuse the same ones that you used for the Update cursor. As before, you'll use the Nothing keyword for the query filter to process every feature and you'll set cursor recycling to True, so that each feature is processed in the same memory location as the one before it.

19 Set the IFeatureCursor variable to create a new feature cursor using the Search method.

```
Set pFCursor = pFClass.Search(Nothing, True)
```

SEARCH VERSUS UPDATE

You use Search cursors to get features, Update cursors to change feature values, and Insert cursors to add new features. Search cursors, however, can also be used to change feature values. With an Update cursor, as you already know, you set an in-memory feature's values with the Value method.

```
pFeature.Value(4) = 25
```

You then use the UpdateFeature method on the cursor to write the value to the table:

```
pFCursor.UpdateFeature pFeature
```

Setting a value with a Search cursor is the same.

```
pFeature.Value(4) = 25
```

However, to write the value to the table with a Search cursor, you use the Store method on the IRow interface (instead of the UpdateFeature method on IFeatureCursor).

```
pFeature.Store
```

These two techniques yield the same result. So what's the difference? The Store method works on individual features, writing each row to the table as its edit is completed. The UpdateFeature method works on the cursor. As edited rows are processed, they are stored in the cursor. When the process ends, all the rows are written to the table together (or in batches of one thousand for large processes).

If you are processing only a handful of features, you can use either method and you won't notice the difference. If you are processing thousands of features, an Update cursor will be faster.

20 Use the feature cursor's NextFeature method to get the first feature.

```
Set pFeature = pFCursor.NextFeature
```

21 Start another Do Until loop.

```
Do Until pFeature Is Nothing

Loop
```

22 Inside the loop, get the stand value of the feature that is being pointed at and add it to a running total of all stand values.

```
dblTotal = dblTotal + pFeature.Value(intPosStandValue)
```

23 Still inside the loop, use the cursor's NextFeature method to get the next feature.

```
Set pFeature = pFCursor.NextFeature
```

By the time the last feature is processed, dblTotal will contain a very large number. (Trees are worth a lot of money, but it also costs a lot to harvest them.) Unfortunately, the dblTotal value won't be attractively formatted. It won't have a dollar sign or commas, and it will show more than two decimal places. It might end up looking something like this: 12345678912.12345.

The CurrencyFormat coclass, one of several ArcObjects classes for formatting numbers, can be useful here. Its INumberFormat interface has a ValueToString method that takes a Double number as its argument. The method adds a dollar sign and commas, shortens the decimal places to two, and returns the formatted number as a string.

24 After the Loop keyword, create a new CurrencyFormat object.

```
Dim pCurrency As INumberFormat
Set pCurrency = New CurrencyFormat
```

25 Use a message box to report the lease value.

```
MsgBox "The lease value is " _
    & pCurrency.ValueToString(dblTotal)
```

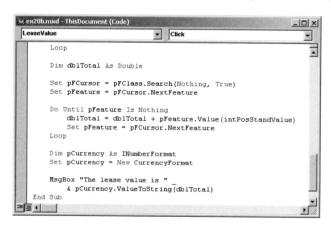

Now you can test the entire LeaseValue code.

26 Close Visual Basic Editor.

27 In the table of contents, right-click the LeaseC layer and click Report Lease Value.

The value is displayed in a message box, so there is no need to open the table.

28 Click OK.

Now try the LeaseD layer.

29 In the table of contents, right-click the LeaseD layer, and click Report Lease Value.

30 Click OK.

As new lease layers come in, the bidders can calculate their value with your button.

Normally, to edit data in ArcGIS you add the Editor toolbar and start an edit session. (Users do this from the interface; programmers do it with various classes, interfaces, properties, and methods.) In this exercise, however, you edited attribute values without starting an edit session. This makes the editing process faster because some of the editing functionality is bypassed.

Edit sessions have some advantages. For example, you can enable the Undo and Redo buttons. If you had used an edit session in this exercise, you could have written code to let the user undo the add field operation or the stand value calculations. Another advantage is that you can offer the user the choice of whether or not to save changes at the end of the session.

On the other hand, there are also some advantages to working outside an edit session. Not starting and stopping an edit session means you have less code to write and the user doesn't have to wait for the session to begin and end. Also, edits are usually processed faster outside an edit session. (This is because, among other things, there is no Undo/Redo information to store.) Working outside an edit session also allows you to code the editing functionality without exposing the user to dialog boxes.

To learn more about editing, refer to the topics *Editor Class* and *Editor Tips* in the developer help. Also, in the "Getting Started" section of the developer help, see the topic *Developing with ArcObjects and COM* and the subtopic *Edit Sessions*.

31 If you want to save your work, click the File menu in ArcMap and click Save As. Navigate to **C:\ArcObjects\Chapter20**. Rename the file **my_ex20b.mxd** and click Save.

As in exercise 20a, you have made changes to a feature class that are independent of the map document. If you want to do this exercise again, you'll need to delete the StandValue fields from the LeaseC and LeaseD layers.

In this book, you've learned the basics of customizing ArcMap. You know how to add buttons and tools to toolbars, menus, and context menus. You can capture user input with input boxes and display information with message boxes. You can make dialog boxes and program them with combo boxes and other controls. You know how to save your changes to a map document or a template.

You know the fundamental concepts of VBA programming. You know about objects, properties, methods, and variables; you know the syntax of writing code. You know what subroutines, functions, events, and property procedures are. You've worked with a number of different event procedures (Click, Initialize, MouseMove, and SelectionChange, to name a few). You know how to call subroutines and functions from event procedures.

You've also learned many important coding techniques: If Then statements, Case statements, Do While and Do Until loops. You know about different kinds of errors and how to debug your code.

From this foundation, you have gone on to learn about ArcObjects. You can navigate object model diagrams and deal with the intricacies of interfaces. You've worked with maps, layers, data sets, feature classes, and features. You can set symbology, make graphics, select features, write queries, update layout elements, and edit data. You know how to put existing ArcGIS functionality into your code with Execute statements. You feel good about yourself when you steal code. You know enough to go ahead on your own.

You've done a lot, but even so, it's just the beginning. (It's no use protesting that you're done with programming. You really can't go back. You'll find yourself idly looking at bits of sample code instead of playing Minesweeper. Then a coworker will ask you to make a tool. One day some people will be standing around a cubicle, trying to figure out why a subroutine won't work, and a voice will say, "The cursor needs to be reinitialized." Much to your surprise, that voice will be your own.)

So whether you are reconciled to your fate or not, sooner or later you'll want to know where to go next. One piece of advice: take it easy. Programming is fun and fruitful when you go in slow, incremental steps. It can be frustrating if you try to do too much too soon.

A good way to get started is to go back to some of the exercises in this book and make modifications to them. Use a different color, a different symbol, a different name. See if you can make a subroutine do a little more than it does, or work with a different type of data; see if you can error-proof it against a certain type of mistake. Look for ways to automate steps that you did manually or that depended on your knowing things in advance about your data. And then?

- Go to the Samples section of the ArcObjects developer help. It contains hundreds of sample procedures arranged by category. Use them, study them, modify them.
- Go to ArcObjects Online (edn.esri.com). You'll find more sample code, technical papers, a discussion forum, and more.
- Join an e-mail discussion list, such as ArcView-L. It's a good way to ask questions and get help. Better yet, start out by viewing the archives of these e-mail discussion lists to see what kinds of questions are asked. Maybe you will find answers to your questions there. To learn how to join a discussion list and view archives, go to support.esri.com/listserve.
- Visit the ESRI Virtual Campus (www.esri.com/training) to find out about live training seminars and workshops on specific ArcObjects topics.
- If you are interested in programming ArcObjects with Visual Basic, .NET, or C++, take one of the ESRI instructor-led courses in advanced ArcObjects component development. To learn more about ESRI instructor-led training, go to www.esri.com/training.
- For the latest information, check the Web site for this book (www.esri.com/ GTKArcObjects).

Data license agreement

Important:
Read carefully before opening the sealed media package

ENVIRONMENTAL SYSTEMS RESEARCH INSTITUTE, INC. (ESRI), IS WILLING TO LICENSE THE ENCLOSED DATA AND RELATED MATERIALS TO YOU ONLY UPON THE CONDITION THAT YOU ACCEPT ALL OF THE TERMS AND CONDITIONS CONTAINED IN THIS LICENSE AGREEMENT. PLEASE READ THE TERMS AND CONDITIONS CAREFULLY BEFORE OPENING THE SEALED MEDIA PACKAGE. BY OPENING THE SEALED MEDIA PACKAGE, YOU ARE INDICATING YOUR ACCEPTANCE OF THE ESRI LICENSE AGREEMENT. IF YOU DO NOT AGREE TO THE TERMS AND CONDITIONS AS STATED, THEN ESRI IS UNWILLING TO LICENSE THE DATA AND RELATED MATERIALS TO YOU. IN SUCH EVENT, YOU SHOULD RETURN THE MEDIA PACKAGE WITH THE SEAL UNBROKEN AND ALL OTHER COMPONENTS TO ESRI.

ESRI License Agreement

This is a license agreement, and not an agreement for sale, between you (Licensee) and Environmental Systems Research Institute, Inc. (ESRI). This ESRI License Agreement (Agreement) gives Licensee certain limited rights to use the data and related materials (Data and Related Materials). All rights not specifically granted in this Agreement are reserved to ESRI and its Licensors.

Reservation of Ownership and Grant of License: ESRI and its Licensors retain exclusive rights, title, and ownership to the copy of the Data and Related Materials licensed under this Agreement and, hereby, grant to Licensee a personal, nonexclusive, nontransferable, royalty-free, worldwide license to use the Data and Related Materials based on the terms and conditions of this Agreement. Licensee agrees to use reasonable effort to protect the Data and Related Materials from unauthorized use, reproduction, distribution, or publication.

Proprietary Rights and Copyright: Licensee acknowledges that the Data and Related Materials are proprietary and confidential property of ESRI and its Licensors and are protected by United States copyright laws and applicable international copyright treaties and/or conventions.

Permitted Uses:

Licensee may install the Data and Related Materials onto permanent storage device(s) for Licensee's own internal use.

Licensee may make only one (1) copy of the original Data and Related Materials for archival purposes during the term of this Agreement unless the right to make additional copies is granted to Licensee in writing by ESRI.

Licensee may internally use the Data and Related Materials provided by ESRI for the stated purpose of GIS training and education.

Uses Not Permitted:

Licensee shall not sell, rent, lease, sublicense, lend, assign, time-share, or transfer, in whole or in part, or provide unlicensed Third Parties access to the Data and Related Materials or portions of the Data and Related Materials, any updates, or Licensee's rights under this Agreement.

Licensee shall not remove or obscure any copyright or trademark notices of ESRI or its Licensors.

Term and Termination: The license granted to Licensee by this Agreement shall commence upon the acceptance of this Agreement and shall continue until such time that Licensee elects in writing to discontinue use of the Data or Related

Materials and terminates this Agreement. The Agreement shall automatically terminate without notice if Licensee fails to comply with any provision of this Agreement. Licensee shall then return to ESRI the Data and Related Materials. The parties hereby agree that all provisions that operate to protect the rights of ESRI and its Licensors shall remain in force should breach occur.

Disclaimer of Warranty: THE DATA AND RELATED MATERIALS CONTAINED HEREIN ARE PROVIDED "AS-IS," WITHOUT WARRANTY OF ANY KIND, EITHER EXPRESS OR IMPLIED, INCLUDING, BUT NOT LIMITED TO, THE IMPLIED WARRANTIES OF MERCHANTABILITY, FITNESS FOR A PARTICULAR PURPOSE, OR NONINFRINGE-MENT. ESRI does not warrant that the Data and Related Materials will meet Licensee's needs or expectations, that the use of the Data and Related Materials will be uninterrupted, or that all nonconformities, defects, or errors can or will be corrected. ESRI is not inviting reliance on the Data or Related Materials for commercial planning or analysis purposes, and Licensee should always check actual data.

Data Disclaimer: The Data used herein for tutorial data, including but not limited to spatial, geographic, map, tabular, statistical, or public-record information, may be actual, fictionalized, and/or a combination thereof. In some cases, ESRI has manipulated, supplemented, and/or applied certain assumptions, analyses, and opinions to the Data solely for educational training purposes. Assumptions, analyses, opinions applied, and actual outcomes may vary. Again, ESRI is not inviting reliance on this Data, and the Licensee should always verify actual Data and exercise their own professional judgment when interpreting any outcomes.

Limitation of Liability: ESRI shall not be liable for direct, indirect, special, incidental, or consequential damages related to Licensee's use of the Data and Related Materials, even if ESRI is advised of the possibility of such damage.

No Implied Waivers: No failure or delay by ESRI or its Licensors in enforcing any right or remedy under this Agreement shall be construed as a waiver of any future or other exercise of such right or remedy by ESRI or its Licensors.

Order for Precedence: Any conflict between the terms of this Agreement and any FAR, DFAR, purchase order, or other terms shall be resolved in favor of the terms expressed in this Agreement, subject to the government's minimum rights unless agreed otherwise.

Export Regulation: Licensee acknowledges that this Agreement and the performance thereof are subject to compliance with any and all applicable United States laws, regulations, or orders relating to the export of data thereto. Licensee agrees to comply with all laws, regulations, and orders of the United States in regard to any export of such technical data.

Severability: If any provision(s) of this Agreement shall be held to be invalid, illegal, or unenforceable by a court or other tribunal of competent jurisdiction, the validity, legality, and enforceability of the remaining provisions shall not in any way be affected or impaired thereby.

Governing Law: This Agreement, entered into in the County of San Bernardino, shall be construed and enforced in accordance with and be governed by the laws of the United States of America and the State of California without reference to conflict of laws principles. The parties hereby consent to the personal jurisdiction of the courts of this county and waive their rights to change venue.

Entire Agreement: The parties agree that this Agreement constitutes the sole and entire agreement of the parties as to the matter set forth herein and supersedes any previous agreements, understandings, and arrangements between the parties relating hereto.

Installing the data

Getting to Know ArcObjects includes one CD that contains the exercise data. The exercise data takes up about 100 megabytes of hard-disk space.

The data installation process takes about five minutes.

Installing the data

Follow the steps below to install the exercise data. Do not copy the files directly from the CD to your hard drive.

1 Put the data CD in your computer's CD drive. In your file browser, click the icon for your CD drive to see the folders on the CD. Double-click the Setup.exe file to begin.

Name	Size	Type	Modified
ArcObjects		File Folder	8/18/2003 1:58 PM
data1.cab	456 KB	WinZip File	8/18/2003 3:35 PM
data1.hdr	46 KB	HDR File	8/18/2003 3:35 PM
data2.cab	41,160 KB	WinZip File	8/18/2003 3:36 PM
IKERNEL.EX_	337 KB	EX_ File	9/4/2001 9:24 PM
layout.bin	1 KB	BIN File	8/18/2003 3:36 PM
SETUP.BMP	709 KB	Bitmap Image	11/6/2002 11:12 AM
Setup.exe	55 KB	Application	9/5/2001 4:23 AM
Setup.ini	1 KB	Configuration Settings	8/18/2003 3:34 PM
setup.inx	134 KB	INX File	8/18/2003 2:16 PM

| 10 object(s) | 41.8 MB | Local intranet |

2 Read the Welcome.

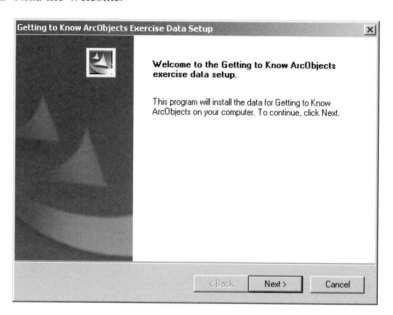

3 Click Next. Accept the default installation folder or navigate to the drive where you want to install the data.

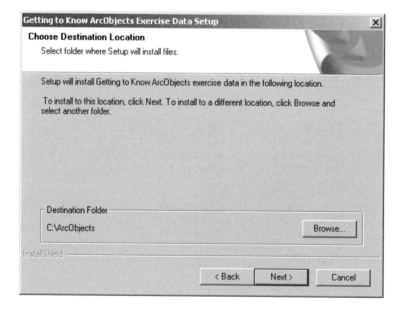

4 Click Next. The exercise data is installed on your computer in a folder called ArcObjects. When the installation is finished, you see the following message:

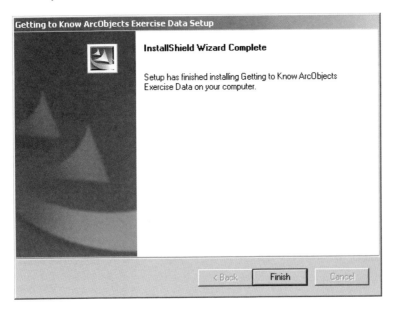

5 Click Finish.

If you have a licensed version of ArcGIS Desktop installed on your computer, you are ready to start *Getting to Know ArcObjects*.

Uninstalling the data

To uninstall the exercise data from your computer, open your operating system's control panel and click the Add/Remove Programs icon. In the Add/Remove Programs Properties dialog, select the following entry and follow the prompts to remove it:

- Getting to Know ArcObjects Exercise Data

Q

Queries. *See* Query statements
Query filters, 332
  creating, 332
  in feature cursors, 348
  and Nothing keyword, 390, 395
  in selection sets, 344
Query statements, 316, 320–22. *See also*
    Query filters
  viewing in Layer Properties, 330
QueryInterface, 159–61, 264
  examples, 163–64, 165–66, 346
  performed by VBA, 214
    examples, 218, 327

R

Rasters, 221. *See also* Layers: adding to
    maps: raster data sets
Recycling cursors, 395–96
Refresh method (IActiveView), 185
Refreshing
  the active view, 184–85
  maps, 322
Removing buttons and tools. *See*
    Commands: removing from toolbars
Renderers, 263, 267. *See also* Layers:
    symbolizing; Symbols
  assigning labels to, 268
  assigning symbols to, 268, 282, 287
  associating with layers, 268, 282
  class breaks, 285
    creating, 286
    properties, setting, 286–87
  creating, 267, 282
RenderMap subroutine, 107
Reporting coordinates. *See* Coordinates,
    reporting
Reset method (IGraphicsContainer), 370
Returned objects. *See* Objects: returned;
    Interfaces: returned
RGB color. *See* Color models
Row buffers, 392
Run button (VBE), 23, 49
Run-time errors. *See* Errors: run-time

S

Sample code, 85, 253, 298. *See also*
    Subroutines: importing
  BufferFeatures, 92–97
  CreateNewChart, 87–90
  CreateOverviewWindow, 82–85
  modifying, 88–89, 93–95
  RenderMap, 107
SaveDocument method (IApplication),
    167
Saving map documents. *See* Map
    documents (.mxd files): saving
Scale dependency. *See* Layers: scale
    dependency, setting
Search cursors. *See* Feature cursors: Search
Search method (IFeatureClass), 348
SelectFeatures method (IFeatureSelection),
    332, 334
Selecting features. *See* Features: selecting
Selection color. *See* Features: selection
    color, setting
Selection sets, 339–41, 342, 343
  creating, 343–45, 345–46
  getting, 342–43
SelectionChange event procedures, 317, 322
SelectionColor property
    (IFeatureSelection), 336
SelectionSet property (IFeatureSelection),
    342
Show method (VBA form), 52, 55
SndPlaySound function, 141
Status bar. *See* ArcMap: status bar
StatusBar property (IApplication), 209
Strings, quoting, 116, 320
Style gallery
  classes, 273
  getting, 275
  items, 273, 275–77
Styles, 274, 278–79. *See also* Style gallery
Subroutines, 77–78. *See also* Procedures
  compared to functions, 77
  creating, 140
  exiting, 365

To resolve problems with the exercise data CD or installation, send an e-mail to ESRI workbook support at *learngis@esri.com*.